Arithmetic Explained

From Memorization to Understanding

Ronald T. Kneusel, Ph.D.

Copyright (C) 2022 by Ronald T. Kneusel. All Rights Reserved.
Correspondence: rkneuselbooks@gmail.com

Contents

1 Numbers   **1**
- 1.1 Numbers as Quantities 1
  - 1.1.1 The Idea of Counting 2
  - 1.1.2 Place Notation 2
  - 1.1.3 Writing Numbers 9
  - 1.1.4 Section Summary 13
- 1.2 Numbers as Distance 14
  - 1.2.1 Introducing the Number Line 15
  - 1.2.2 Moving to the Left 16
  - 1.2.3 Integers 19
  - 1.2.4 Section Summary 20
- 1.3 Chapter Summary 21
- 1.4 Terms and Concepts 21
- 1.5 Exercises 24

2 Addition   **27**
- 2.1 Addition By Grouping 27
  - 2.1.1 Grouping and Place Notation 28
  - 2.1.2 Section Summary 30
- 2.2 Addition By Moving 30
  - 2.2.1 Adding Two Single-Digit Numbers 31
  - 2.2.2 Single-Digit Addition Facts 35
  - 2.2.3 Section Summary 37
- 2.3 Adding Two Two-Digit Numbers 37
  - 2.3.1 Section Summary 40
- 2.4 Adding Two Numbers of Any Size 40
  - 2.4.1 Section Summary 43
- 2.5 Adding Many Numbers 43
  - 2.5.1 Section Summary 47
- 2.6 Chapter Summary 48

iv   CONTENTS

    2.7 Terms and Concepts 48
    2.8 Exercises 49

3 Subtraction   **51**
    3.1 Subtraction By Taking Away 51
          3.1.1 Section Summary 53
    3.2 Subtraction By Moving 54
          3.2.1 Single-Digit Subtraction Facts 56
          3.2.2 Section Summary 57
    3.3 Subtraction Without Borrowing 58
          3.3.1 Section Summary 60
    3.4 Subtraction With Borrowing 60
          3.4.1 Repeated Borrowing 64
          3.4.2 The Trouble With Zero 65
          3.4.3 Section Summary 68
    3.5 Subtracting a Larger Number From A Smaller Number 68
          3.5.1 Section Summary 69
    3.6 Chapter Summary 70
    3.7 Terms And Concepts 71
    3.8 Exercises 71

4 Mixed-Sign Numbers   **75**
    4.1 Mixing It Up with the Number Line 76
          4.1.1 Addition 76
          4.1.2 Subtraction 78
          4.1.3 Section Summary 80
    4.2 Addition and Subtraction Free-for-All 81
          4.2.1 A Cheat Sheet 85
          4.2.2 Section Summary 86
    4.3 Chapter Summary 86
    4.4 Terms And Concepts 87
    4.5 Exercises 87

5 Multiplication   **89**
    5.1 From Addition to Multiplication 89
          5.1.1 Section Summary 92
    5.2 The Multiplication Table 92
          5.2.1 Options for Learning the Multiplication Table . 97
          5.2.2 Section Summary 101
    5.3 Multiplying By Single-Digit Numbers 101
          5.3.1 Section Summary 104
    5.4 Multiplying Two-Digit Numbers 104

		5.4.1 Something About Ten 107
		5.4.2 Section Summary 108
	5.5	Multiplying Many-Digit Numbers 108
		5.5.1 Section Summary 112
	5.6	What About Negative Numbers? 112
		5.6.1 Why Negative If One Number Is Negative? . . . 113
		5.6.2 Why Positive If Both Numbers Are Negative? . 114
		5.6.3 Section Summary 116
	5.7	Chapter Summary 116
	5.8	Terms And Concepts 117
	5.9	Exercises 117

6 Division   121

6.1	Division As Repeated Subtraction 122
	6.1.1 A Bit Of Theory (Optional) 124
	6.1.2 Section Summary 125
6.2	The Essence of Division 125
	6.2.1 Section Summary 129
6.3	Dividing By a Single-Digit Number 129
	6.3.1 The Recipe 135
	6.3.2 An Alternative Viewpoint 136
	6.3.3 Danger, Will Robinson! 136
	6.3.4 Section Summary 138
6.4	Will It Divide? 138
	6.4.1 Section Summary 141
6.5	What's My Sign? 141
6.6	Chapter Summary 142
6.7	Terms and Concepts 143
6.8	Exercises 144

7 More Division   147

7.1	Two-Digit Divisors 147
	7.1.1 Section Summary 152
7.2	Arbitrary Divisors 153
	7.2.1 Section Summary 156
7.3	Chapter Summary 156
7.4	Exercises 157

8 Fractions   159

8.1	Fractions Are... 160
	8.1.1 Parts of A Whole 160
	8.1.2 On The Number Line 162

- 8.1.3 Rational (Optional) 164
- 8.1.4 Section Summary 165
- 8.2 Fractions Undercover 166
  - 8.2.1 Fractions Have Aliases 166
  - 8.2.2 Reducing Fractions 167
  - 8.2.3 Section Summary 171
- 8.3 Dividing Fractions 171
  - 8.3.1 Reciprocals 171
  - 8.3.2 Dividing By Multiplying 173
  - 8.3.3 Section Summary 174
- 8.4 Fractions With Like Denominators 174
- 8.5 Fractions With Unlike Denominators 175
  - 8.5.1 The Lazy 176
  - 8.5.2 Section Summary 177
- 8.6 Ratios and Proportions 178
  - 8.6.1 Ratios 178
  - 8.6.2 Proportions 180
  - 8.6.3 Section Summary 182
- 8.7 Negative Fractions 183
  - 8.7.1 Section Summary 184
- 8.8 Mixed Numbers and Improper Fractions (Oh, My!) . . 184
  - 8.8.1 Mixed Numbers 185
  - 8.8.2 Mixed Numbers Are Improper Fractions in Disguise 187
  - 8.8.3 Section Summary 189
- 8.9 Chapter Summary 189
- 8.10 Terms and Concepts 190
- 8.11 Exercises 192

9 Powers 195
- 9.1 Powers and Exponents 195
  - 9.1.1 Powers of Fractions 198
  - 9.1.2 What About Negative Bases? 200
  - 9.1.3 Section Summary 201
- 9.2 More Trouble With Zero 201
- 9.3 Squares and Square Roots 202
  - 9.3.1 Square Roots of Negative Numbers 204
  - 9.3.2 Section Summary 205
- 9.4 Negative Exponents 205
- 9.5 Multiplication and Division with Powers 206
  - 9.5.1 Section Summary 208
- 9.6 Number Bases 208

CONTENTS   vii

      9.6.1 Section Summary 211
  9.7 Chapter Summary 212
  9.8 Terms and Concepts 212
  9.9 Exercises 214

10 Decimals   217
  10.1 Concerning Decimals 217
      10.1.1 The Fractional Part 218
      10.1.2 Speaking Decimals 222
      10.1.3 On the Number Line 222
      10.1.4 Section Summary 224
  10.2 Adding and Subtracting Decimals 225
      10.2.1 Section Summary 228
  10.3 Comparing Decimal Numbers 228
  10.4 Multiplying Decimals 229
      10.4.1 Multiplying by Powers of Ten 232
      10.4.2 Section Summary 233
  10.5 Converting Fractions to Decimals 233
      10.5.1 Section Summary 240
  10.6 Dividing Decimals 240
      10.6.1 Dividing by Powers of Ten 241
      10.6.2 Why The Multiplication Algorithm Works 243
      10.6.3 Dividing by Integers 244
      10.6.4 Dividing by Decimals 246
      10.6.5 Section Summary 248
  10.7 Decimal Numbers and Computers 248
  10.8 Chapter Summary 250
  10.9 Terms and Concepts 251
  10.10 Exercises 251

11 Percents   255
  11.1 What Is A Percent? 256
  11.2 Fractions, Decimals, and Percents 256
      11.2.1 Fractions to Percents 257
      11.2.2 Decimals to Percents 259
      11.2.3 Section Summary 260
  11.3 Percent of a Number 260
      11.3.1 X is What Percent of Y? 264
      11.3.2 Section Summary 265
  11.4 Percents in Action 266
      11.4.1 Recognizing Percents 266
      11.4.2 What's The Tip? 269

11.4.3 Poverty Rates 270
11.4.4 A Tale of Two AIs 271
11.4.5 Compound Interest 274
11.4.6 It's On Sale! 275
11.4.7 Section Summary 277
11.5 Percent Change 277
11.5.1 Lemonade Stand 279
11.5.2 Population Growth 282
11.5.3 Percentage Points 284
11.5.4 Section Summary 285
11.6 Chapter Summary 285
11.7 Terms and Concepts 286
11.8 Exercises 287

12 Epilogue, Resources, And Cheat Sheet   **291**
12.1 Epilogue 291
12.2 Resources 291
12.3 Arithmetic Cheat Sheet 293

13 Solutions to Exercises   **303**

Introduction

Arithmetic concerns itself with the addition, subtraction, multiplication, and division of numbers. If we think of things like birds, rocks, red Swingline staplers, etc., then addition is the merging of two groups of things to form a larger group of things. Likewise, subtraction takes a group and removes some portion of it to leave a smaller group. Multiplication is repeated addition. Division is repeated subtraction. These are the "big four" operations of arithmetic.

This book explains arithmetic beginning with the idea of a number and what it means to represent it. Once we know what numbers are and how we want to represent them, we can define how to add, subtract, multiply, and divide them. That's the primary goal of this book.

Along the way, we'll learn that there are different kinds of numbers: positive numbers, negative numbers, whole numbers, integers, fractions, mixed numbers, and decimals. We'll explore how to add, subtract, multiply, and divide all of these. If you find numbers confusing or strange, don't worry, numbers do make sense, and by walking through the book together, I'm sure you'll come to see that this is so. It's all good.

Arithmetic is sometimes labeled "elementary mathematics" as if that were somehow a negative. In this context, the word "elementary," means "elemental" – something that is at the core, something foundational. How could it be that the foundation of a topic is anything other than critically important? Without arithmetic, mathematics becomes, at best, extremely difficult, if not impossible.

In truth, there are many professional mathematicians, we call them "number theorists," who spend their entire careers working with numbers like 1, 2, 3, 4, etc. That's it. The numbers are rich enough that understanding them and their relationships with each other is enough to occupy the mind for a lifetime. Elementary mathematics, i.e., arithmetic, is utterly foundational and critically important. And, lucky for us, also seriously fun and understandable.

Who Is This Book For?

This book is for lots of people. I might even be so bold as to say that this book is for everyone. Why? Because arithmetic is the most practical of all mathematics in that all of us need what it offers from time to time, regardless of who we are or what we do.

To be more specific, this book is for people who want to gain a deeper understanding of arithmetic. Most of us learned arithmetic as children. That's all well and good, but children, for the most part, only learn the mechanics without a deeper understanding of why. For example, to understand *why* multiplication works the way it does, or *why* long division does what it does, you need to be older, more mature, and able to think more abstractly.

If you plowed through arithmetic in elementary school, then promptly ignored it while preoccupied with the essential things in life, that's perfectly fine. You are reading this paragraph, so you see a reason to "brush up" on arithmetic. You are not alone, not by a long shot. "Use it or lose it" goes the old saying. And it's true, even with things like arithmetic.

Perhaps you never had the opportunity to learn arithmetic well, regardless of the reason. If so, then this book is definitely for you. The mere fact that you can read this sentence means you have all that is needed to understand arithmetic, and by understanding arithmetic, to opening the door to a much larger mathematical world. It's no secret that our modern age depends critically on information. Information is usually expressed as numbers, and to manipulate numbers, you need arithmetic.

What Can You Expect to Learn?

By the end of this book, you will know how to read, write, and say numbers. You will know how to add, subtract, multiply, and divide any kind of number. You will know what fractions are, and percents, and how to work with them in everyday settings. And, most importantly, I hope you will have come to understand that mathematics can be fun and deeply satisfying.

What I Expect You to Know Already

I expect you to know how to read this sentence. That's it. We'll cover everything else together.

Road Map

"Grasshopper was tired. He lay down in a soft place. He knew that in the morning the road would still be there, taking him on and on to wherever he wanted to go." - Arnold Lobel, *Grasshopper on the Road*

Like grasshopper, we're on a journey, one we hope never ends, nor should we want it to. However, even a wanderer wants a map from time to time. Here's ours.

It's best to read the book straight through. Some of the material will undoubtedly be familiar, but I encourage you to read those chapters and sections anyway. It's often helpful to review, even things we think we know well. I do it often. Besides, reading will help you get to know me and how I present information. And, the better you know my approach, the more you'll get from the book.

In general, each chapter consists of multiple sections, with subsections and so on when warranted. In addition, many sections have a "Section Summary," and each chapter concludes with an overall "Chapter Summary."

After the chapter summary is a list of terms and concepts. These reflect what was introduced in the chapter, and, at times, add a bit more.

Exercises come next. Unlike many books, here there are only a few exercises, at least by type. Sometimes there are multiple problems per exercise, but all are of the same kind. My goal with the book's exercises isn't to wear you down. Do as many or as few as you feel you need. The bulk of the available exercises are on the book's website; see below.

Each chapter includes a "Think About It" section designed to broaden your perspective regarding the chapter's topics. I think you'll enjoy reading through them. I certainly enjoyed writing them. You'll find my comments on these sections in the appendix.

Finally, here's the road map:

1. Numbers   The book begins at the beginning asking "what is a number?" Answering that question isn't as straightforward as it might appear at first. Our struggle with the idea of number will lead us to a deeper appreciation of the power and utility of our current number system: place notation. I don't expect it to be superseded anytime soon, though I might argue at some point for a change of base. Don't worry; you'll eventually understand what I mean, if you don't already.

2. **Addition**   Addition is the easiest of the big four operations, so we begin with it. The chapter introduces addition as grouping together and as motion along a number line. The number line is a valuable tool, one that we use consistently throughout the book.

3. **Subtraction**   The opposite of addition, of grouping together, is taking away, which we call "subtraction." This chapter builds on addition and teaches us the meaning and mechanics of subtraction.

4. **Mixed-Sign Numbers**   Some arithmetic books shy away from the concept of negative numbers. We won't. The number line shows us how to deal with negative numbers, and we encounter negative numbers in real life, too, especially during the winter if you live in the North or when your bank reminds you, ever so kindly, that you've spent more money than you actually have. This chapter helps us understand how to apply addition and subtraction when one or both of the numbers is negative.

5. **Multiplication**   This chapter introduces multiplication as repeated addition, then builds the (often dreaded) multiplication table. Once we have the multiplication table, we tear it apart to see that a handful of simple rules reduces the table to a manageable size. With the multiplication table out of the way, we dive into the multiplication algorithm, the recipe we use to multiply two numbers. I go beyond the method and explain the meaning of the madness.

6. **Division**   If multiplication is repeated addition, then division is repeated subtraction. Division is the most complex of the big four operations, but by now, we're up to the challenge. As with multiplication, I'll go beyond the *how* and explain the *why* as well. Knowing why makes it easier to see and remember the how.

7. More Division   Division needs another chapter. This one is it.

8. Fractions   Fractions are mysterious beasts, masters of disguise. In this chapter, we explore fractions and learn how to work with them in all their guises. For regular numbers, addition and subtraction are straightforward, while multiplication and division take some effort. For fractions, the reverse is true.

9. Powers   In arithmetic, multiplying a number by itself repeatedly is known as raising the number to a power. This brief chapter introduces the idea of powers, and the related term, exponent, both of which are needed for Chapter 10.

10. Decimals   Decimal numbers are numbers that combine a whole number with a fraction but use place notation to write both parts. This chapter introduces decimal numbers, along with what they mean when written and where to find them on the number line. Then, we learn how each of the big four operations works when we have decimal numbers. Fortunately, little needs to change, so working with decimals isn't much of a stretch.

11. Percents   Fractions typically live between 0 and 1 on the number line. Fractions can be written as decimal numbers, but it's inconvenient to work with numbers like 0.17, so people since ancient times have written such numbers as fractions with a denominator of 100, i.e., by the hundred, so they can talk about 17 percent instead of 0.17. Don't worry if none of that makes sense. Making sense of it is precisely what this chapter sets out to do.

12. Epilogue, Resources, And Cheat Sheet   No book covers everything in sufficient detail or in a way that resonates with everyone. Therefore, I would be remiss if I didn't offer you additional resources, places you can go to clarify what might still be unclear or to move beyond arithmetic altogether. That's the point of this final chapter. Oh, there's also a quick summary of the mechanics of arithmetic, something to look back at from time to time as needed.

Appendix A contains solutions for the exercises and my comments on the "Think About It" sections.

Worksheets

To learn math, you must do math. Regardless of the level, every student of mathematics has had the experience of reading something or hearing a lecture where the topic seems crystal clear, only to find when attempting to use the topic on their own that it's now as clear as mud. The antidote for this phenomenon is to *do* mathematics. Fight with the concepts until you own them. To win the fight, you need exercises, problems with solutions so you can check your work.

Most math books include lots of exercises at the end of each chapter or section. As I indicated above, this book is a little different. While there are a few exercises at the end of each chapter, with corresponding solutions at the back of the book, most of the exercises are not in the book, but instead available on a companion website,

```
https://github.com/rkneusel9/ArithmeticExplained
```

This site contains example problems with solutions organized by chapter. My hope is that you'll print the pages and work on the page directly. If you can't print the pages, copy the problems carefully and work on them that way. Solutions to each problem are given.

If you have trouble with a problem or a question about the book, please let me know at `rkneuselbooks@gmail.com`. I look forward to hearing from you.

Chapter 1

Numbers

What is a number? Mathematicians have a formal definition of *number*, but that definition, while fascinating, isn't particularly helpful for day-to-day use. In this chapter, we'll contemplate the idea of a number in two different ways. First, we'll think about a number as a *quantity*, how many of something. This is the most natural way to think about numbers. Second, we'll think about a number as a *distance*, how far we are from some origin point. Thinking of a number as a distance isn't immediately apparent, but, as we'll learn, it's a valuable way to think about numbers because thinking about numbers as distance introduces us to the *number line*, a tool we'll rely on heavily in later chapters.

1.1 Numbers as Quantities

How old are you? How long have you lived in this city? How many children do you have? The answer to each of these questions involves a quantity of something, either years or children. How many eggs are in a dozen? Again, the answer is a quantity, something *discrete* that can be counted. Usually, the word *many* is used for something that can be counted, as opposed to the word *much*, which is used for something that can't easily be counted, like how much sand there is on a particular beach. We'll stick with "many" phrases for now; "much" will appear in later chapters.

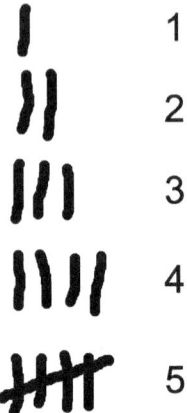

Figure 1.1: Counting with tally marks

1.1.1 The Idea of Counting

Counting is so natural that we seldom think about it. After all, it's been around a long time: ancient humans counted by making tally marks on animal bones. We still use tally marks, usually in groups of five, as in Figure 1.1.

In the figure, the tally marks are on the left, first one, then two, three, and four. The fifth tally mark is a slash across four marks. Using the slash separates groups of five for faster counting. We'll come back to that later in the book as well.

Tally marks are used to record numbers in a nonverbal way. As people developed writing, they needed some way to record numbers, the quantity of something. Recording how many of this and how many of that, be it what someone owned or owed to someone else, was a strong motivation for the development of writing in the first place. The vast majority of clay tablets from ancient Mesopotamia, where writing was first developed, concern just that: who owns how many of this and who owes who how many of that. It took centuries before anyone thought of using writing for something other than record-keeping.

1.1.2 Place Notation

Look again at Figure 1.1. On the left are the tally marks, and on the right are the numbers we recognize. How did humanity get from

1.1. NUMBERS AS QUANTITIES

tally marks to the numbers we use now? I'm not talking specifically about history, but conceptually, the mental process that changed tally marks into the *Arabic numerals* or *digits* that we use today.

Let's count to ten with tally marks,

|
| |

Counting to one hundred this way would be quite annoying; we'll see an example below. Therefore, it would be nice if we had a simpler way to indicate in writing which number we mean. We already have words like *one* and *two*, so we could use those,

one
two
three
four
five
six
seven
eight
nine
ten

but writing the whole word is tedious as well. What if, instead, we invent some way to mean specific numbers, like five and ten? Let's use letters for one, five, and ten: I, V, and X. Now we can count like this,

I
II
III
IIII
V
VI
VII
VIII

VIIII
X

Instead of five ones, we use V. For six we use VI because six is one more than five. We can get fancy and replace IIII with IV, since four is one before five. Likewise, we could use IX in place of VIIII since nine is one before ten.

As you might know, the number system we just invented has a name, *Roman numerals*. It's the same system the ancient Romans used. They learned it from the ancient Egyptians. In the end, though, it's just a shorthand way to use tallies with some groups given a special meaning, like V for five.

It is entirely possible to write numbers this way. The Roman Empire was quite successful for centuries. But, it isn't convenient, and, as it turns out, there is a better way to write numbers: *place notation*. Let's see how it works.

Humans have ten fingers on both hands. Let's use this fact to count, where we'll use a special symbol for each number. On the one hand, we have five fingers,

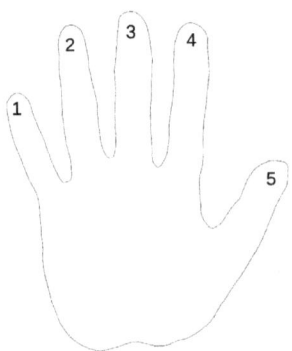

The special symbols run from the left to the right: 1, 2, 3, 4, and 5. The symbols don't just label fingers; they also tell us how many fingers. If three fingers, we use 3 because when 3 is written above, three fingers have been numbered.

Let's keep going and label the right hand as well,

1.1. NUMBERS AS QUANTITIES

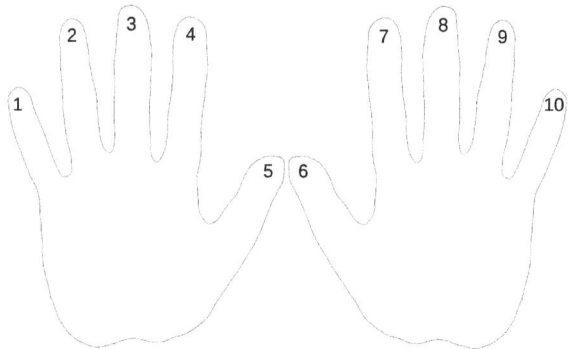

We have unique symbols for all the fingers on both hands. Or, have we? Look at the little finger of the right hand, the rightmost finger. It has two symbols on it: 10. We recognize the first symbol, 1, as the symbol we used for one. What's the second symbol, the 0? Here's where things get interesting. Here's where we make a conceptual leap that took humanity centuries to take.

We want some way of writing a symbol for the word *ten*. We might invent a symbol for ten as we did for all the other fingers, but instead, we used two symbols: 1 and the new symbol, 0. The new symbol is zero, a symbol that means none. If I ask you "how many T. rex dinosaurs live in zoos?" your answer would be "zero" as they are all, sadly, extinct. So, 0 means none, nothing; there aren't any. If that's so, then what's 0 doing labeling the tenth finger? It's a finger, after all, not nothing.

To understand 10, we need to learn the idea behind place notation. As the name suggests, it's a notation, a way for writing numbers. And, it uses places, which I'll describe shortly.

Because we have ten fingers, when we count, we use symbols for all the fingers until we get to the tenth one. To mark ten, we add a new symbol to the end of the symbol for one; we add zero. We do this because each *place* in the sequence of symbols for a number has a meaning—hence "place notation."

The symbols 1 through 9 are markers for how many. What do they mark? For a number with two digits, they mark the number of tens and the number of ones. We read from left to right, so we put the mark for the tens before the mark for the ones: 10 means one ten and zero ones. We now see why we need a symbol for none. We need zero to indicate there are no ones.

If 10 is ten, what is 11? It's one ten and one one, and ten plus one is eleven, so 11 is eleven. Likewise, 12 is one ten and two ones, which is twelve. Using place notation, we can write any two-digit number by breaking it up into how many tens and how many ones. The ancient Romans and Egyptians understood to use special symbols for specific numbers to make grouping easier, but they did not make the leap to the *order* in which a small set of symbols can be written to specify a number. In place notation, *where* the symbol appears tells us how many of that group. For two-digit numbers, that means either tens or ones. We'll get to larger numbers in a bit.

Let's write some numbers along with what they literally mean using words,

14	one ten and four ones
25	two tens and five ones
39	three tens and nine ones
64	six tens and four ones
78	seven tens and eight ones
99	nine tens and nine ones

In each case, the first symbol, a digit, tells us the number of tens and the second digit the number of ones. Notice, what the number means is sometimes different from how we *say* the number when speaking.

To better understand place notation, it's helpful to work through the act of counting. Above, with the hands, we counted from one to ten. Let's count again, but this time to twenty-four running first down the left side then the right,

1	13
2	14
3	15
4	16
5	17
6	18
7	19
8	**20**
9	21
10	22
11	23
12	24

Notice how 10 and 20 are in bold. The zeros show up at ten and twenty because those numbers have rolled over in a sense; they are a specific number of tens with no extra to fill in the ones, so we put a zero in the ones.

1.1. NUMBERS AS QUANTITIES

Notice, after ten, counting to twenty repeats the digits in order for the ones until nineteen, 19. To add one more to get to twenty, we repeat what we did for nine to ten: we move the tens to the next digit and set the ones to zero. Then, after every ten, we roll over to increment the tens digit and set the ones digit back to zero to repeat the sequence.

If we do this long enough, we'll get to the number ninety-nine, 99. We have a nine in the one's place, so our rule says to increment the tens and make the ones a zero. But, the tens is a nine, and, applying the same rule, we should increment something and make the nine in the tens place a zero. What do we increment? We increment the next digit place to the *left* to create a three-digit number: one hundred, 100.

In a three-digit number, the first digit is now counting the number of hundreds. The second digit counts the number of tens, and the third digit the number of ones.

As we count, the next number is formed by moving the ones place to the next digit, say 7 to 8, while keeping all the other digits the same until the ones place has a 9. In that case, the next number puts a 0 in the ones place and moves the tens place to the next digit. If the tens place has a 9, it is replaced by a 0 as well, and the next place to the left, the hundreds, moves to the next digit. This recipe repeats forever. Every time the leftmost digit is a 9 and needs to increment, just add a new digit to the left and make all the other digits zero.

Consider these examples,

```
      1            91
      2            92
      3            93
      4            94
      5            95
      6            96
      7            97
      8            98
      9            99
     10           100
     11           101
     ▲▲           ▲▲▲
      │ ones        │ ones
      tens         tens
                   hundreds
```

On the left, we count from one to eleven. I labeled the places, the ones and the tens. Likewise, on the right, I'm showing counting from ninety-nine, 99, to one hundred, 100. Again, the places are labeled. Note, the places are often referred to as *columns*. So, the first column on the right is the one's column, the second the ten's column, and so on.

Here's a question: what comes after 999? According to the recipe, the nine in the one's column becomes a zero, and the tens column is incremented. But, the tens column is a nine, so it also becomes a zero, and the hundreds column is incremented. Since the hundreds column is nine, it likewise becomes zero, and we add a new digit, a one, on the left. So, the number after 999 is 1000. The fourth place has a name: the thousands. Therefore, 1000 is the number one thousand because there is a one in the thousands place and zeros in the hundreds, tens, and ones places.

Let's get a feel for how big numbers are when written in place notation. We'll use tally marks. For example, here's the difference between 1 and 2,

$$| \quad \text{versus} \quad ||$$

Not much to write home about. Here's the difference between 10 and 20,

$$\begin{array}{c} |\,|\,|\,|\,| \\ |\,|\,|\,|\,| \end{array} \quad \text{versus} \quad \begin{array}{c} |\,|\,|\,|\,| \\ |\,|\,|\,|\,| \\ |\,|\,|\,|\,| \\ |\,|\,|\,|\,| \end{array}$$

Finally, here's the difference between 100 and 200,

[10 rows of 10 tally marks] versus [10 rows of 20 tally marks]

With place notation, writing 10 or 100 involves adding one more zero, but what the two numbers *mean* in terms of quantity is very different. The number 100 is much larger than 10.

1.1. NUMBERS AS QUANTITIES

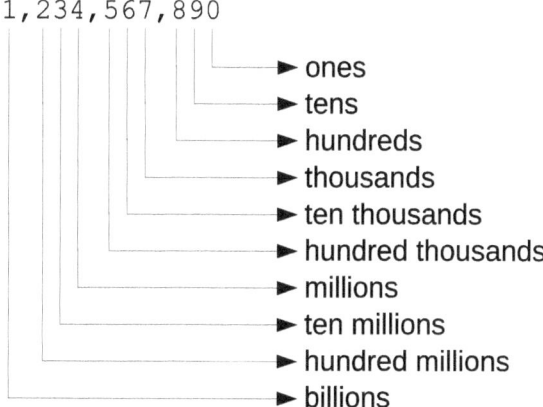

Figure 1.2: Place names

There is no limit to how large a number we can write using place notation. We can always add a new digit on the left when necessary. Therefore, not all of the places have names associated with them. However, many do. For example, consider Figure 1.2.

In Figure 1.2, the commas are not part of the number. Rather, they serve to separate the digits into groups of three to make it easier to read.

The number in Figure 1.2 has ten digits. It's a big number. How big? The place-name for the leftmost digit is billions. There's a one in there, so the number is at least one billion.

1.1.3 Writing Numbers

We can write numbers using digits without speaking them out loud, but at times we need to say the numbers to communicate with other people. As it happens, writing a number in word form is the same as speaking the number out loud. Therefore, if we learn to write numbers in word form, we'll learn how to say numbers as well.

We already know how to write single-digit numbers, just write the digit. Let's learn how to write two-digit numbers followed by three-digit numbers. After three digits, the process gets easier.

Writing Two-Digit Numbers

Two-digit numbers from ten to twenty have special names. That's because humans are, well, human, and not always logical. Let's write the numbers and their names,

> 10   ten
> 11   eleven
> 12   twelve
> 13   thirteen
> 14   fourteen
> 15   fifteen
> 16   sixteen
> 17   seventeen
> 18   eighteen
> 19   nineteen
> 20   twenty

We have no choice but to memorize these names, though there is the remnant of an old pattern for 13 through 19, the "teens." "Teen" comes from Old English and means "ten more than," so fourteen, for example, is ten more than four.

Once we reach twenty, things become more consistent. We need only name the group, meaning the tens, followed by a digit name,

> 21   twenty-one
> 22   twenty-two
> 23   twenty-three
> 24   twenty-four
> 25   twenty-five
> 26   twenty-six
> 27   twenty-seven
> 28   twenty-eight
> 29   twenty-nine

Notice that in word form, the number is hyphenated and that each number is the group, here twenty, followed by the digit name. This pattern repeats for all the groups up to one hundred. Therefore, we need only know the group names,

1.1. NUMBERS AS QUANTITIES

20	twenty
30	thirty
40	forty
50	fifty
60	sixty
70	seventy
80	eighty
90	ninety

We now know all we need to name any two-digit number. For example,

24	twenty-four
33	thirty-three
42	forty-two
18	eighteen
57	fifty-seven
66	sixty-six
78	seventy-eight
84	eighty-four
99	ninety-nine

Let's move on to three-digit numbers.

Writing Three-Digit Numbers

Numbers less than one hundred use special forms. Fortunately, the hundreds portion of three-digit numbers is straightforward: the hundreds portion is simply the number followed by the word *hundred*. For example,

100	one hundred
200	two hundred
300	three hundred
400	four hundred
500	five hundred
600	six hundred
700	seven hundred
800	eight hundred
900	nine hundred

Notice, there is no hyphen as with two-digit numbers and that we say "hundred," not "hundreds." To name any three-digit number, we write the hundreds followed by the two-digit name. For example,

123	one hundred twenty-three
315	three hundred fifteen
351	three hundred fifty-one
555	five hundred fifty-five
768	seven hundred sixty-eight
999	nine hundred ninety-nine

When writing three-digit numbers, don't use commas between words and don't use the word *and*. We'll use "and" later in the book when we work with other kinds of numbers.

Writing Numbers with More Than Three Digits

We saw in the previous section that each place has a name: the ones, tens, hundreds, thousands, etc. Look again at Figure 1.2. After the thousands place comes the ten thousands place then the hundred thousands. The three places, thousands, ten thousands, and hundred thousands, form what's called the thousands *period*. Likewise, the ones, tens, and hundreds places form the ones period. The period names are,

$$\underbrace{314}_{\text{trillions}},\underbrace{159}_{\text{billions}},\underbrace{265}_{\text{millions}},\underbrace{358}_{\text{thousands}},\underbrace{979}_{\text{hundreds}}$$

If you compare the period names with the column names in Figure 1.2, you'll see that the period names match the name of the first column in the period. To name a number larger than three digits, we say the number in each period as a three-digit number followed by the period name. Consider these examples,

1,234	one thousand two hundred thirty-four
23,453	twenty-three thousand four hundred fifty-three
852,587	eight hundred fifty-two thousand five hundred eighty-seven
111,222,333	one hundred eleven million two hundred twenty-two thousand three hundred thirty-three

There are several things to note in these examples. First, the numbers are aligned on the right, meaning the ones column of each

1.1. NUMBERS AS QUANTITIES

number is aligned. We'll see this again when we study addition. Next, the numbers use commas to group digits in sets of three where we count from right to left to place commas: 23,453, not 234,53.

Since we write numbers in word form as we speak them, try saying the numbers above. If you're like me, you'll be tempted to put the word *and* in at various places. Resist the temptation.

Writing large numbers in words can be tedious. Consider,

314,159,265,358,979   three hundred fourteen trillion one hundred fifty-nine billion two hundred sixty-five million three hundred fifty-eight thousand nine hundred seventy-nine

We don't need to do it often, at least.

How should we handle zeros in the number? Remember, we use zero in place notation to indicate that the number doesn't have any of that particular column. We'll see this explicitly later in the book when we discuss addition. For now, consider these examples,

1,003,001	one million three thousand one
700,000	seven hundred thousand
1,000,000	one million
40,040,040	forty million forty thousand forty
1,000,000,002	one billion two

We're most tempted to use "and" when zeros appear in the number, but, to be consistent, we shouldn't. That's why the first example is "one million three thousand one" not "one million three thousand *and* one." Also, the second and third examples clarify that when the rest of the number is zero, we don't say it.

1.1.4 Section Summary

In this section, we explored the idea of number as quantity, how many of something. We learned that tally marks can track numbers to record counting but that the utility of tally marks is limited. We then learned about place notation, the method we use to write numbers. In place notation, each place (or column) has a value. We use special symbols, the digits 1 through 9 and 0, to mark how many of each place go into the number.

Next, we looked at writing numbers in word form to match how we say them out loud. The smaller numbers have special forms, like "seventeen" and "fifty-three," but larger numbers become consistent, if tedious at times. Finally, we saw that zeros in a number had little effect on how we write the number in words, except when all of the digits to the right of a particular place are zero, as in "28,000", which is "twenty-eight thousand."

It's natural to think of number as quantity, and, very often, a number is a quantity, the measurement or count of something. However, for the remainder of this chapter, we'll mix things up and contemplate number as distance.

1.2 Numbers as Distance

In the previous section, we contemplated numbers as quantities. When early humans counted, likely using their fingers, they counted like this,

$$1, 2, 3, 4, 5, 6, \ldots$$

The line above is one way to present a *mathematical series*, a set of numbers that use a rule to create the next number in the series. For counting, the rule is: the next number is the previous number plus one. The dots, also called an *ellipsis*, indicates that the series continues, in this case, forever.

Notice, we started counting from one. This is a natural thing to do; in fact, mathematicians call the counting numbers beginning with one the *natural numbers*. When they talk about the *set* of natural numbers, they use a special symbol: ℕ.

The word *set* was in italics in the previous paragraph for a reason. In math, a set is a collection of unique things; in this case, the set ℕ is the collection of all natural numbers. It's not essential to remember much about sets, but I call your attention to them because we'll encounter many other sets of numbers in this book, and it's good to know how to refer to them and, later, to see the relationships between them.

So, counting starts with one. This fits nicely with our discussion of number as quantity, of how many. Let's make a slight change to how we count. Let's start with zero instead,

$$0, 1, 2, 3, 4, 5, 6, \ldots$$

The collection of numbers when counting from zero is often called the *whole numbers*. Mathematicians, being human beings, sometimes use different definitions for things, and some mathematicians

1.2. NUMBERS AS DISTANCE

Figure 1.3: A number line

will call the whole numbers, including zero, the natural numbers. I find it easier to split hairs and call counting from one the natural numbers and adding zero into the mix the whole numbers. There is no formal mathematical symbol that refers to the whole numbers. We'll use W and call it good.

Counting from zero, check. But, why? If numbers are how many of something, then we can have zero of that thing. For example, how many alien starships are in orbit around the moon? Zero, sadly. But let's look at this another way. You and your friend live in Milwaukee, Wisconsin. Your friend asks you how far it is from Milwaukee to Minneapolis, Minnesota. You are a geography whiz so you know it's 297 miles as the crow flies. Your friend, impressed with your knowledge, now asks far it is from Milwaukee to Chicago, Illinois? 83 miles, you declare triumphantly. With a smirk, your friend then asks how far it is from Milwaukee to Milwaukee. Without missing a beat, you say 0 miles. Why zero? Because Milwaukee is your origin point, it's where you measure distances *from*. This is the core idea behind thinking of number as distance: that there is an *origin*, a place to call zero, and numbers name how far we are from the zero point.

1.2.1 Introducing the Number Line

Consider Figure 1.3.

Figure 1.3 shows what looks a bit like a ruler. There's a line with an arrow to the right and a set of vertical hash marks at equal distances from each other. The hash marks are labeled beginning with zero. This is a *number line*, and it's a useful way to think about numbers and the things we can do with numbers, like add and subtract them.

The number line, because it starts with zero, shows us whole numbers, the set of numbers we're calling W. In reality, the number line shows a whole lot more, but for now, concentrate on the hash marks and the numbers below them.

Since we will make heavy use of the number line in later chapters, let's spend a bit of time getting used to it. One thing we'll do with the number line is to mark specific numbers. For that, we'll use circles. For example, here's how we mark the numbers 1, 3, and 7,

How does the number line indicate distance? The numbers tell us how far, meaning how many numbers there are, between zero and the number we're talking about. For example, on the number line above, we marked 1, 3, and 7. Place your finger on zero. Now, move it over one space at a time until you get to the 7, counting as you go. When your finger hits 7, your count should also be 7. Therefore, the distance between 0 and 7 is 7.

Sometimes, we'll place markers or arrows above the number line. Here, I'm showing the whole numbers with the natural numbers below them,

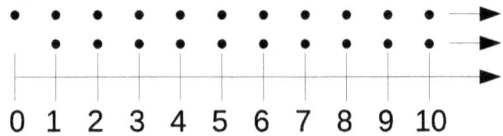

The arrows indicate that the numbers continue forever to the right. Remember, the whole numbers include zero, while the natural numbers do not.

1.2.2 Moving to the Left

I said that this is a number line,

And it is. However, the arrow goes from zero to the right only. Lines can go in either direction. Let's extend the line to the left as well,

1.2. NUMBERS AS DISTANCE

It sure looks like there should be something to the left of zero. We have hash marks and an arrow. How should we label the hash marks?

We're thinking of number as distance, and we saw that counting from zero to a position on the number line tells us the distance from zero. Place your finger on zero and move it four spaces to the left. How far are you from zero? You're at the fourth hash mark, just as you would be if counting from zero and moving to the right. Therefore, we might be tempted to label the hash mark on the left 4.

In a sense, that wouldn't be wrong, we are a distance of four from zero, but it would also be confusing. To indicate which 4 we're referring to we need to supply more information. We might say "4 to the left" to distinguish it from "4 to the right," but that's cumbersome. So, we'll simply put a "-" in front of the numbers on the left instead. Doing so gives us a new number line,

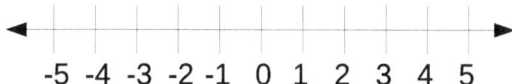

To indicate a distance of four to the right, we use 4, and to mean four to the left, we use -4.

The "-" has a name, it's a *minus sign* or, sometimes, a *negative sign*. You'll most often see "minus sign," but, as I'll explain in a moment, that can be confusing, so I'll stick with "negative sign" instead.

Why might "minus sign" be confusing? Because the word *minus* is also used to mean subtraction. In fact, "-" *is* the symbol for subtraction, as we'll see in later chapters. In reality, however, marking a number with a negative sign to mean it's to the left of zero on the number line is quite different from subtraction, so it's an unfortunate historical fact that the same symbol is used for both. To further complicate matters, putting "-" in front of something, like a number, really means *negation*, making it the opposite. That's how mathematicians use "-" in this case, -4 is the negation of 4.

We call the numbers to the left of zero *negative numbers*. And, we call the numbers to the right of zero *positive numbers*. Negative numbers have a negative sign in front of them. Positive numbers, on rare occasions, might have a "+" in front of them. This is a *plus sign*. We won't use plus signs to indicate positive numbers in this book. If there is no negative sign, the number is positive.

By definition, if a number is to the left of another on the number line, we say that number is *less than* the other. So, 1 is less than 4 because 1 is to the left of 4 on the number line. By this definition, all the negative numbers are less than zero because they are all to the left of zero on the number line.

If numbers are only quantities, there is no such thing as a negative number; you can't have less than zero of something. However, numbers are not only quantities. If numbers are thought of as distances from zero, we *can* have less than zero, and those are the numbers we call negative.

We run across negative numbers in everyday life. If you live where it gets cold in the winter, the weather report might say that the high temperature today will be -5 F, negative five degrees Fahrenheit.

The sentence above warrants two comments. First, it presents a practical example of number as distance, here the distance from an arbitrary mark on the thermometer labeled zero where -5 is *below* zero because, traditionally, analog thermometers are oriented vertically.

The second comment is more subtle. The weather report said the high temperature for the day would be -5. That means all other temperatures will be less than -5, meaning they are *more negative* than -5, or, to the left of -5. The phrase *more negative* is typically used to mean one negative number is smaller than another negative number.

Mathematicians use a special symbol to mean one number is less than another, <. For example, to write that 2 is less than 5, we write $2 < 5$. Similarly, to say that -4 is less than -2 we write $-4 < -2$. On the number line, we show this with arrows,

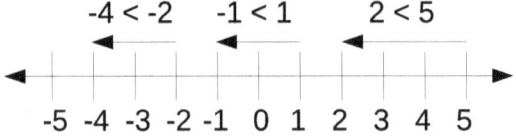

Here the line is to the left to mean "less than," and the mathematical way of writing the meaning of the line is above it. Notice, $-1 < 1$

1.2. NUMBERS AS DISTANCE

crosses the zero mark. That's okay.

If a number to the left of another is less than the other number, then a number to the *right* of another is *greater than* the first. Mathematicians use > to mean one number is greater than another. So, −2 > −4 and 5 > 2. Using the number line and arrows gives us,

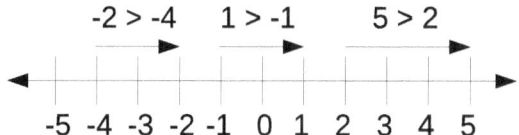

The arrows now point to the right to say one number is larger than another. Notice, the relationships shown in this number line and the one above using less than are the same; they indicate the same relationship using different perspectives.

If a number is to the right of another on the number line, that number is larger, regardless of sign. So, moving further to the right means the numbers get larger. Moving further to the left means the numbers get smaller, even if they are more negative and further to the left of zero. A graphic might help here,

The important point is that "larger" means to the right, always, while "more negative" and "more positive" means further from zero to the left or the right, respectively.

1.2.3 Integers

Above, we discussed the set natural numbers, the ones mathematicians denote ℕ—all the positive numbers beginning with one. We then added zero to the set and called that the whole numbers, 𝕎.

Does the set of numbers including all the negative numbers, zero, and all the positive numbers have a special name? It does; these numbers are known as *integers*. Mathematicians use ℤ to

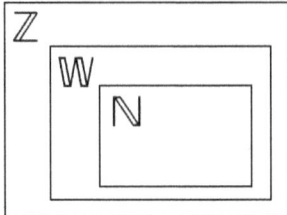

Figure 1.4: A diagram showing the relationship between ℕ, 𝕎, and ℤ.

refer to the integers. Why ℤ and not 𝕀? Because not all mathematicians speak English. ℤ is from *zahlen*, the German word for numbers. Many mathematicians have built highly successful careers working with nothing but integers.

Mathematicians like to group and categorize things, usually as sets. With the integers, we now have three sets of numbers we can talk about. There are more, and we'll get to them in time. The sets have relationships between them that we can express with a simple diagram, see Figure 1.4.

Diagrams like this have a name. They are called *Venn diagrams* after John Venn, the English mathematician who developed them in the 19th century. A Venn diagram shows relationships between sets of things. For example, if a box is entirely inside another box, then the set represented by the smaller box is in some way "inside" the larger set. In this case, the diagram tells us that all the natural numbers fit in the set of whole numbers because the whole numbers include all natural numbers and zero. Likewise, all the whole numbers, and therefore all the natural numbers, fit inside the integers. We'll encounter Venn diagrams in later chapters when we discuss other kinds of numbers.

1.2.4 Section Summary

This section introduced us to the number line, a fundamental tool we'll frequently use in later chapters. We learned what the number line is, how to represent numbers on it, and what happens if we extend the number line to the left of zero: negative numbers. We used the negative numbers as a stepping stone to the ideas of greater than or less than, which we illustrated using the number line.

We concluded the section by introducing an essential set of numbers, the integers. We saw how the integers include all the whole numbers and all the natural numbers. We then learned a bit about Venn diagrams to see, visually, the relationship between integers, whole numbers, and natural numbers. Finally, we ended with a promise: there are other types of numbers to come.

1.3 Chapter Summary

The main goal of this chapter was to think about number in two complementary ways: as quantity and as distance. For the former, we started by counting, first with tally marks, then Roman numerals, and finally with digits, which led us to place notation. We spent a fair bit of time working with numbers, especially how to write them in word form, thereby learning how place notation works and how to say numbers when we read them. We also compared the size of numbers by using tally marks to show that a slight change in the number of digits in a number can imply a significant difference in the meaning of that number.

The second section of the chapter introduced the number line and negative numbers, the numbers to the left of zero. Along the way, we learned about greater than and less than as a means for arranging numbers or talking about which is bigger or smaller than another. We showed these relationships on the number line as well. Finally, we learned about three different sets or groupings of numbers: natural numbers (counting numbers), whole numbers, and the most important number type of all, the integers.

With numbers now part of our toolkit, we are ready to move on to the first of the four basic arithmetic operations, addition.

1.4 Terms and Concepts

We introduced the following terms and concepts in this chapter.

Arabic Numerals The symbols for the numbers one through nine along with zero.

Counting Beginning usually with one, the act of producing a sequence of numbers by increasing the previous number by one.

Digit A numeral, $0, 1, 2, 3, \ldots, 9$, used in writing a number. A digit occupies one place in the number.

Ellipsis A symbol, ..., that means something continues. In mathematics, the ellipsis is usually used to indicate that a series continues where the rule for generating the series is given explicitly or easily derived from the numbers before the ellipsis.

Greater Than The symbol >. The idea of one thing being bigger than another. For numbers, a first number is greater than a second number if the first number is to the right of the second number on the number line.

Integer All the numbers we've marked on the number line in this chapter. The integers, denoted $\mathbb{Z}$, include all negative numbers, zero, and all positive numbers.

Less Than The symbol <. The idea of one thing being smaller than another. For numbers, a first number is less than a second number if the first number is to the left of the second number on the number line.

Minus Sign The symbol -. In some cases, - means subtraction. In other cases, - means negation, a negative number.

More Negative The idea of a negative number being further to the left on the number line than another negative number.

Natural Number The collection, or set, of all numbers beginning with one. All the counting numbers, $1, 2, 3, \ldots$.

Negation The act of negating something. For numbers, negation is changing 4 to -4, for example. Similarly, the negation of -4 is 4.

Negative Number Any number to the left of zero on the number line.

Negative Sign Another name for "Minus Sign" with the explicit meaning of negation or marking a number as negative.

Number (as Distance) The concept of number as a measure of how far from zero on the number line.

Number (as Quantity) The concept of number as how many of something.

Number Line A line with hash marks indicating where zero and other numbers are, similar to a ruler. The number line is a useful tool for expressing the relationship between numbers.

1.4. TERMS AND CONCEPTS

Period A set of three digits grouped when speaking or writing numbers using commas. Periods separate the digits to make reading the number easier and to clarify how the number should be spoken or written.

Place Notation The concept of expressing numbers using digits and positions where the order of the digits indicates how many of each group go into making up the number. For example, 123 is one hundred, two tens, and three ones.

Positive Number Any number to the right of zero on the number line.

Roman Numerals The number system used in ancient Rome. A simple evolution of tally marks.

Series A collection of numbers generated by a rule. Counting is a series where the rule is "the next number is one larger than the previous number."

Set A collection of something, like a collection of numbers. The natural numbers are a set. The whole numbers are set, as are the integers.

Symbol A mark (glyph) used to indicate something else. For numbers, the Arabic numerals are the symbols used to write numbers using place notation.

Tally Marks An ancient method for counting using slash marks, one per whatever is being counted.

Venn Diagram A diagram showing relationships between sets of things. For numbers, a Venn diagram can show that all natural numbers are part of the whole numbers and all whole numbers are part of the integers.

Whole Number All positive numbers and zero.

Zero As a quantity, zero means there is none of something. Zero is the origin point of the number line, the place from which all other numbers as distance are measured. In place notation, zero indicates that the number doesn't have any of that place; 103 is one hundred and three ones, no tens.

1.5 Exercises

Exercise 1

Write the following numbers in word form.

783
10,056
57,899
111,111
1,030,005
123,456,789

Exercise 2

Write the following in number form.

- seventeen
- eight thousand eighty-seven
- forty-two thousand three hundred sixty-four
- ten million ten thousand ten
- two hundred ninety-nine million seven hundred ninety-two thousand four hundred fifty-eight

Exercise 3

Label each expression below as either *true* or *false*. Use this number line to help you,

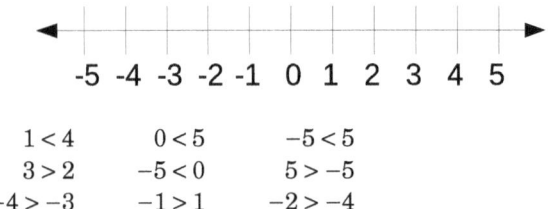

1 < 4   0 < 5   −5 < 5
3 > 2   −5 < 0   5 > −5
−4 > −3   −1 > 1   −2 > −4

Challenge 1

In this chapter, we learned that the ancient Romans used I, V, and X to represent groups of one, five, and ten, respectively. To make larger numbers, the Romans utilized L for 50, C for 100, D for 500, and M for 1,000. With that knowledge, see if you can write the following numbers using Roman numerals. As noted in the chapter, the Romans wrote one smaller unit before a larger one when possible, so not IIII but IV, not VIIII but IX. Similarly, 40 is XL, 90 is XC, and 900 is CM, one hundred before one thousand. Typically, the larger groups come before the smaller, so 21 is XXI, not IXX.

    42
   127
  1066
  1941
  2000
  2022

Think About It

We use place notation counting by tens, first ten ones, then tens, then hundreds, and so on. There is nothing special about ten other than humans have ten fingers on both hands, and hands are the most obvious part of the body to use for counting. What if we only had three fingers on each hand? How might place notation work in that case?

Chapter 2

Addition

Addition is the first of the four fundamental arithmetic operations. It's also the most straightforward. By the end of this chapter, you'll understand addition in two ways. In the previous chapter, we contemplated number as both quantity and distance. In this chapter, we'll work with addition from both of those viewpoints. Addition is grouping together; this matches the idea of number as quantity. Addition is also moving along the number line, meaning addition is also related to distance.

Talking about addition in both of these ways helps cement the idea of addition; it explains the *what*: what addition is and what it represents. With that background, we then move forward and learn the *how*: the *algorithm*, or recipe, allowing us to add two or more numbers.

Let's begin.

2.1 Addition By Grouping

Addition can be thought of as taking two groups and merging them. Let's examine an addition problem as two groups merged, then we'll talk about it and the notation it introduces,

$$3 + 4 = 7$$

Multiple things are going on in the figure above. Let's start with the dots on the top line. On the left, there are two groups of dots. The first group has three dots, and the second has four dots. On the right, there are seven dots, which we see is what we get by imagining the group of three pushed next to the group of four. Therefore, three *plus* four *equals* seven. Here, *plus* is another word for addition, for adding things together. And, *equals* is a word that, for now, we'll use to mean the answer, what you get when you add two numbers together.

The figure has a second line, $3 + 4 = 7$. This is *math notation*, the mathematical way to indicate that three plus four equals seven. Here, + is an *operator*, the symbol for addition, the *plus sign*. Likewise, = is the symbol for an *equals sign*, the result of some mathematical operation, like addition. In reality, = means that two things are the same; the thing on the left of the = is the same as the thing on the right.

The second line in the figure above is also a *mathematical equation*, a statement of a fact. In this case, the fact states that if you take a group of three things and add a group of four more things to it, you'll end up with a group of seven things.

Another way to write $3 + 4 = 7$ is vertically,

$$\begin{array}{r} 3 \\ +\,4 \\ \hline 7 \end{array}$$

We line up the numbers we want to add above the horizontal line and write the answer, the *sum*, below. We'll use the vertical approach most often. It's handy when working out the solution to an addition problem. For the time being, however, remember that we are talking about addition as grouping things together; we're not focused on adding just yet, only on what addition means.

2.1.1 Grouping and Place Notation

If we have two groups of something, like black dots, and merge them, we've added them together. We just saw this above. Adding in this way works regardless of the numbers involved. For example,

2.1. ADDITION BY GROUPING

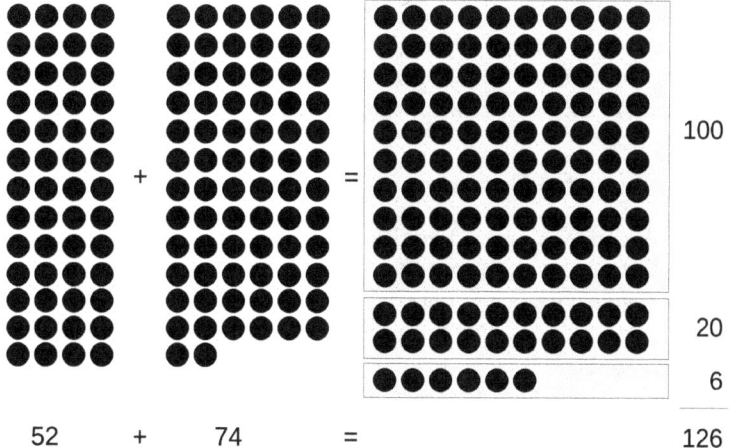

Here we're using dots and groups to show that $52 + 74 = 126$, or in other words, that the sum of fifty-two and seventy-four is one hundred twenty-six. On the left are two groups, one of fifty-two dots and another of seventy-four dots. Merging them gives us the one hundred and twenty-six dots on the right. I'll explain the gray boxes in a bit. Vertically, we write this problem as,

$$\begin{array}{r} 52 \\ + 74 \\ \hline 126 \end{array}$$

There are one hundred and twenty-six dots to the right of the equals sign. I drew some gray boxes around some of the dots. The largest covers one hundred, the next largest covers twenty, and the smallest covers six. I wrote numbers indicating how many dots to the right: 100, 20, and 6.

Below these numbers, I wrote 126, which I say is the answer when adding 52 and 74. The vertical equation above shows this as well. So, why the boxes? The goal here is to see that the number 126 is, as we discussed in Chapter 1, one group of a hundred plus two groups of ten plus six ones. These are precisely the number of dots in the gray boxes above. Therefore, we can write, even though we haven't yet discussed how to add it, that

$$\begin{array}{r} 100 \\ 20 \\ + 6 \\ \hline 126 \end{array}$$

For now, keep this equation in mind, the idea of splitting a number into the sum of each of the digits in their positions. This is how a group of something is viewed as a number in place notation. We'll come back to this idea later in the chapter.

2.1.2 Section Summary

In this section, we saw that addition could be viewed as the grouping of collections or sets of things together. Three dots grouped with four dots leaves us with seven dots, therefore, $3 + 4 = 7$. We also learned that place notation splits the number of dots into groups by ones, tens, hundreds, and so on to see how to go from a collection of dots to a number representing the collection. This is viewing addition as merging groups of things, addition as quantity.

2.2 Addition By Moving

In Chapter 1, we discussed the idea of number as a distance from the zero mark on a number line,

In this section, we'll learn that addition can be thought of as moving from one number on the number line to another. Therefore, addition is *motion* along the number line.

We'll do this by learning how to add two single-digit numbers together. We'll clarify the idea with examples to help us remember the basic addition facts.

How does the number line help us understand addition? Let's find out. In this example, I'm showing $3 + 4 = 7$,

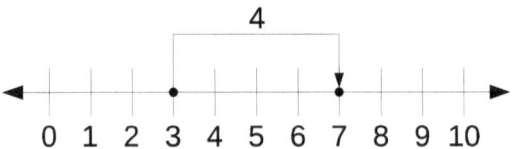

We begin at 3 on the number line, where the dot is. To *add* 4, we *move* from 3 four spaces *to the right*. Doing so gets us to 7, which is

the answer, as we saw above. Therefore, moving a certain number of spaces to the right from a starting position is addition.

Let's see another example, here $2 + 8 = 10$,

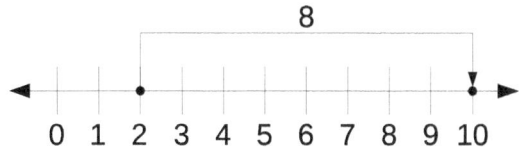

We start at 2, then move 8 spaces *to the right* to land at 10, which is the correct answer. I'm emphasizing *to the right* because we'll learn in another chapter that moving to the left has meaning. You might be able to guess it already.

Let's repeat it: moving right a set number of spaces on the number line from a starting number to an ending number is addition. However, notice something useful about addition. Above I show the number line for $3 + 4 = 7$. Here's the number line showing $4 + 3$,

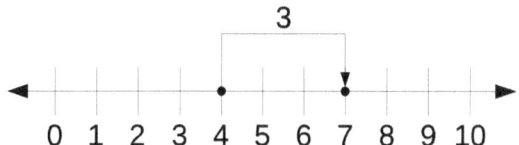

We land at 7. Therefore, $4 + 3 = 7$. However, $3 + 4 = 7$ as well, so we might, thinking of the equals sign as showing that both sides are the same thing, write $3 + 4 = 4 + 3$. For addition, the *order* in which we add the two numbers does not matter; the answer is the same. Mathematicians use the word *commutative* for this situation. What's essential for us to remember is that addition works regardless of the order of the numbers. We'll learn in Chapter 5 that multiplication works this way as well.

Notice, also, that I wrote $3 + 4 = 4 + 3$ above using the equals sign to mean the left and the right are the same. What I find helpful is to imagine a see-saw with the "=" in the middle. Since the two sides are the same, the see-saw will be level, as it would be if two children who weigh the same were sitting on each end.

2.2.1 Adding Two Single-Digit Numbers

Consider this rather elaborate number line,

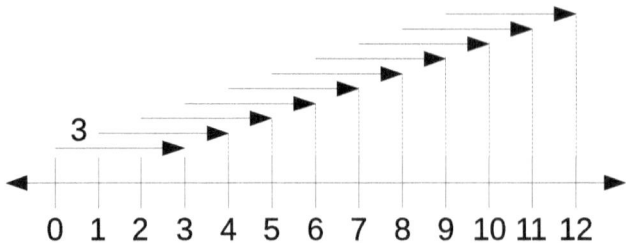

We have the numbers from zero through to twelve. Above the numbers are ten arrows, all the same length and all facing to the right. Each arrow begins above one number and ends at another. Then there is a 3 floating above the first arrow. What's going on here? Let me explain.

This number line shows us how to add three to each of the numbers from zero through nine: we begin at the first number, move three to the right, and end at the answer. Using math notation, the number line is telling us that the following are true,

$$0 + 3 = 3$$
$$1 + 3 = 4$$
$$2 + 3 = 5$$
$$3 + 3 = 6$$
$$4 + 3 = 7$$
$$5 + 3 = 8$$
$$6 + 3 = 9$$
$$7 + 3 = 10$$
$$8 + 3 = 11$$
$$9 + 3 = 12$$

Now, consider this number line,

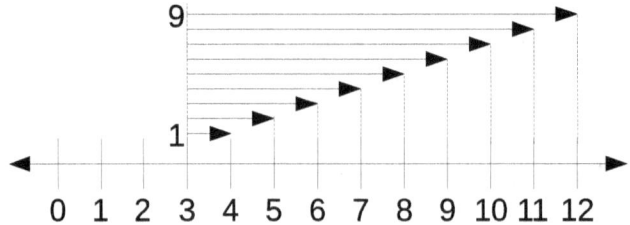

2.2. ADDITION BY MOVING 33

Here, every arrow begins at 3 but the arrow lengths change from one to nine, so, from three, first move one, then two, then three, and so on until nine. This number line tells us these are also true,

$$3 + 1 = 4$$
$$3 + 2 = 5$$
$$3 + 3 = 6$$
$$3 + 4 = 7$$
$$3 + 5 = 8$$
$$3 + 6 = 9$$
$$3 + 7 = 10$$
$$3 + 8 = 11$$
$$3 + 9 = 12$$

Even though the two number lines look different, we now know that $3 + 2 = 2 + 3$ and so on, therefore, both number lines express the same set of facts, they tell us how to add single-digit numbers involving three.

Another thing to notice is that the first number line started at zero to tell us that $0 + 3 = 3$. The second number line doesn't show us this fact, but it should be clear that starting at three and not moving (that's zero motion) leaves us at three, so $3 + 0 = 3$ is also true.

We could, if we wanted, make similar number lines for each of the digits, zero through nine. We won't; instead, we'll see a different way to show the addition facts below.

Before we move on, though, the two number lines above contain one more item to comment on, and it's crucially important if we want to use place notation successfully. Look at these facts,

$$7 + 3 = 10$$
$$8 + 3 = 11$$
$$9 + 3 = 12$$

They do not look like the other addition facts involving three. They have two digits in their answers. It's important to realize the distinction between the answer and how we write it. From the number lines, we found the answer in each case the same way: we moved from the starting number three spaces to the right. However, by doing that, we ended up with a final number that is too large to express with a single digit, as those digits only cover the numbers from zero through nine. In particular, to write the answer to $7+3$ using place notation, we write 10 because we end on the number ten, which is one ten and no ones.

When this happens, the single-digit answer we had before rolls over, as it were; we end up with an answer too large for a single digit, we have to increment the next digit, just like we did in Chapter 1 when counting, which is really just adding one repeatedly.

Adding three to seven produces a *carry*, a new digit one place to the left, which must be added to the next column, in this case, the tens column. And, as with counting, if there is no column already to the left of the column we're working with, we add the carry to zero, which is the same as just putting the 1 in the next column over. This explains how in $8+3=11$, the first one is the carry, the increment to the tens column, and the second one is the one remaining. Both together are eleven, which is one ten and one one.

When adding two numbers, we can only ever get a carry of one. We'll see later in the chapter that larger carries are possible when adding more than two numbers at once.

It might help to see the addition vertically, which is how we'll work through problems later in this chapter,

$$\begin{array}{r} \mathbf{1} \leftarrow \text{carry} \\ 7 \\ +\ 3 \\ \hline 1\mathbf{0} \end{array}$$

Here, I show the carry, the small, bold 1, above the tens column. I also show the zero in the answer in bold. The idea is to read the small carry digit and the zero as 10. The carry digit is added to any other tens digits, but, as there aren't any, we might imagine it falls down the column into the answer.

Using this convention, $9+3=12$ is written as,

$$\begin{array}{r} 1 \\ 9 \\ +\ 3 \\ \hline 12 \end{array}$$

where I have removed the bold digits, and the "carry" label. As we work with addition problems in this chapter, we'll stick with this convention of showing the carry as a small digit above the numbers being added. This is how most books show a carry.

2.2. ADDITION BY MOVING

Table 2.1: The single-digit addition facts

+	0	1	2	3	4	5	6	7	8	9
0	0	1	2	3	4	5	6	7	8	9
1	1	2	3	4	5	6	7	8	9	10
2	2	3	4	5	6	7	8	9	10	11
3	3	4	5	6	7	8	9	10	11	12
4	4	5	6	7	8	9	10	11	12	13
5	5	6	7	8	9	10	11	12	13	14
6	6	7	8	9	10	11	12	13	14	15
7	7	8	9	10	11	12	13	14	15	16
8	8	9	10	11	12	13	14	15	16	17
9	9	10	11	12	13	14	15	16	17	18

The critical point is that whenever the value we want to place in a column is greater than nine, we have a carry that moves over to the next column on the left.

2.2.2 Single-Digit Addition Facts

Above, I showed number lines that expressed all the single-digit addition facts for the number three. We can always use a number line to work out an addition fact, especially for single-digit numbers, but we can also present the facts as a table. Once we know these facts, or are quick to work them out if we don't remember them, we can move on to adding more complicated pairs of numbers and, eventually, an arbitrary set of numbers.

Table 2.1 shows the single-digit addition facts.

Let's learn how to use the table to find the sum of seven and eight, $7 + 8$. First, find the row beginning with seven. Then, slide your finger along that row until you get to the column labeled 8. The number under your finger is 15, which is the answer, $7 + 8 = 15$. Notice, if you start at row eight and move over to column seven, you also end up with 15, $8 + 7 = 15$. Finally, the table works the same if you start at a column and slide down to a row. Beginning at column 8 and sliding down to row 7 still gives you 15 as the answer. The table lets us quickly find the sum of any two single-digit numbers, from $0 + 0 = 0$ to $9 + 9 = 18$.

The table also shows us some interesting patterns. For example, look at the diagonal from the upper right to the lower left. Every entry is 9. Moving through the table along this diagonal shows us all the ways to add two numbers and get nine as the answer,

$$0 + 9 = 9$$
$$1 + 8 = 9$$
$$2 + 7 = 9$$
$$3 + 6 = 9$$
$$4 + 5 = 9$$

All the flipped versions like $9 + 0$ and $7 + 2$ similarly add to nine. Notice the pattern: as the first digit goes up by one, the second digit goes down by one so that the sum always remains nine. Thinking of addition as quantity, we might imagine that we're moving a black dot from the second group to the first group. We still get the same answer. For example,

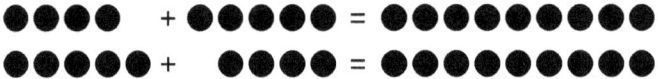

Patterns like this show up in the table for all the diagonals from the upper right to the lower left. A particularly useful set of patterns are those that add to ten,

$$1 + 9 = 10$$
$$2 + 8 = 10$$
$$3 + 7 = 10$$
$$4 + 6 = 10$$
$$5 + 5 = 10$$

When adding multiple numbers at a time, it helps to look for those pairs that add to ten.

Study Table 2.1 for a bit. Do you see any other patterns? What about the diagonals from the upper left to the lower right? Or how the values change from one column to the next? A lot of interesting mathematics has come from studying the patterns that show up in numbers, especially when arranged according to some rule, like addition. We'll see a table much like Table 2.1 when we study multiplication.

2.2.3 Section Summary

This section presented the idea of addition as motion along the number line from a starting number to an ending number where the distance traveled is the number added. From this idea, we explored adding single-digit numbers, which led us to the concept of a carry and, ultimately, Table 2.1, the table of single-digit addition facts. We'll use the table to learn how to add two two-digit numbers. As we do, we'll eventually memorize the table, so we need not refer to it so often.

2.3 Adding Two Two-Digit Numbers

We are now ready to work through the addition algorithm, the recipe that lets us add numbers together. I could have started the chapter here, but to do so would have been unfair to you. One of my goals with this book is to help you understand *why*. Above, we considered addition in the same way we considered numbers to get at the *idea* of addition, not just how to do it via a recipe. But now is the time to, as the ads say, "just do it."

Let's add two random numbers: 24 and 35. To add the numbers, write them vertically,

$$\begin{array}{r} 24 \\ + 35 \\ \hline \end{array}$$

Notice, the numbers line up *on the right* so that the ones columns match. It's essential to do this, so the columns line up, the ones of the first number with the ones of the second, the tens of the first with the tens of the second, and so on.

The algorithm consists of repeated single-digit additions from right to left. If there is a carry, we mark it above the next column to the left of the current column and add it in when working with that column. Therefore, to add 24 and 35, we begin by adding the ones, 4+5, which is 9. We write the 9 under the horizontal line like so,

$$\begin{array}{r} 24 \\ +35 \\ \hline 9 \\ \uparrow \end{array}$$

Remember that this isn't the complete answer; we're not yet finished. When we are, I'll make the entire answer **bold** so you know we are done. Also, I added an arrow below the 9 to indicate the column we just added.

The ones column has been added. We now move one column to the left and add the tens column, so adding 2+3, we get 5,

$$\begin{array}{r} 2\,4 \\ +3\,5 \\ \hline \mathbf{5\,9} \\ \uparrow \end{array}$$

The full answer is **59**, fifty-nine. That's it. That's the addition of two two-digit numbers. Begin on the right, add the digits, mark any carry on the tens column, then move to the left one column and repeat adding in any carry.

Let's try another example, 56 and 38. You might want to pause here and try it yourself on a piece of paper. Do refer back to Table 2.1 as needed. I strongly encourage you to *write* the answer on paper, don't try to do it in your head.

First, we write the problem vertically, then we add the ones column, 6+8. Notice, 6+8=14 is more than 9, so we have to do a carry. Therefore, write the 4 in the answer for the ones column and carry the 1 to the tens,

$$\begin{array}{r} 1 \quad \leftarrow \text{carry} \\ 5\,6 \\ +3\,8 \\ \hline 4 \\ \uparrow \end{array}$$

To finish, we add the tens and include the carry. Therefore, we add $5+3$ to get 8 and then add one more for the carry, $8+1=9$.

$$\begin{array}{r} 1 \quad \leftarrow \text{carry} \\ 5\,6 \\ +3\,8 \\ \hline \mathbf{9\,4} \\ \uparrow \end{array}$$

We have our answer, $56+38=94$, ninety-four.

Let's work a few more examples before moving on to adding arbitrary pairs of numbers. For this example, let's add 89 and 93. As

2.3. ADDING TWO TWO-DIGIT NUMBERS

before, we start with the ones: $9+3 = 12$. As $12 > 9$, twelve is greater than nine, we have a carry, so we write the 2 under the ones column and the carry above the tens,

$$\begin{array}{r} 1 \leftarrow \text{carry} \\ 8\,9 \\ +9\,3 \\ \hline 2 \\ \uparrow \end{array}$$

Now we add the tens column remembering to include the carry, $8 + 9 = 17$ and $17 + 1 = 18$. Okay, the result is also larger than nine, so, what do we do? We do exactly as we did for the ones: we write the 8 in the tens column of the answer and the carry in the next column to the left, the hundreds,

$$\begin{array}{r} 1 \leftarrow \text{carry} \\ 8\,9 \\ +9\,3 \\ \hline 8\,2 \\ \uparrow \end{array}$$

Notice, the carry is above the hundreds column. Neither 89 nor 93 have digits in that column. We can imagine the hundreds columns have zeros in them, there are no hundreds in these numbers. We then add $0 + 0 = 0$ and then the carry, $0 + 1 = 1$, to get a 1 in the hundreds column,

$$\begin{array}{r} 1 \leftarrow \text{carry} \\ 8\,9 \\ +9\,3 \\ \hline \mathbf{1\,8\,2} \\ \uparrow \end{array}$$

In practice, when we add the leftmost digits, we write the result even if it's greater than nine. Notice, the first two digits of the answer above are 18, exactly what we found when adding the tens columns, including the carry from the ones. I like to imagine that the carry has fallen down into the hundreds column of the answer because there's nothing beneath it.

2.3.1 Section Summary

In this section, we learned how to add two two-digit numbers. First, add the digits of the ones column writing the sum in the ones column of the answer. If the sum is greater than nine, write the ones column of the sum with the tens column of the sum as the carry above the tens column. Then, add the digits of the tens column, including any carry. Write that sum in the tens column of the answer, even if there is a carry, which becomes the hundreds column of the answer.

2.4 Adding Two Numbers of Any Size

We're now ready to add two numbers of any size, meaning any number of digits. First, write the two numbers vertically, making sure to line up the ones columns on the right as we did above. Then, work from right to left, adding column by column moving any carries to the next column on the left.

Let's add 254,567,599 and 3,435. First, we'll write the problem vertically lining up the ones columns,

$$\begin{array}{r} 254567599 \\ +3435 \\ \hline \end{array}$$

We now proceed column by column from the right beginning with $9 + 5 = 14$, which has a carry,

$$\begin{array}{r} 1 \\ 254567599 \\ +3435 \\ \hline 4 \\ \uparrow \end{array}$$

Moving on to the tens column gives us $9 + 3 = 12$ and $12 + 1 = 13$,

$$\begin{array}{r} 11 \\ 254567599 \\ +3435 \\ \hline 34 \\ \uparrow \end{array}$$

Next comes the hundreds column, $5 + 4 = 9$ and $9 + 1 = 10$,

2.4. ADDING TWO NUMBERS OF ANY SIZE 41

$$\begin{array}{r} 111\\ 254567599\\ +3435\\ \hline 034\\ \uparrow \end{array}$$

We're almost done. We have the thousands column next, $7 + 3 = 10$ and $10 + 1 = 11$,

$$\begin{array}{r} 1111\\ 254567599\\ +3435\\ \hline 1034\\ \uparrow \end{array}$$

What should we do with the ten thousands column? There are no more digits to the smaller number, 3,435, so the digits above "fall down" into the answer. Really, the missing leading digits of 3,435 to match the corresponding digits of the larger number are all zero and adding zero to a number doesn't change it. However, we do need to add the carry above the 6 in the ten thousands column to get 7. Therefore, the final answer is,

$$\begin{array}{r} 1111\\ 254567599\\ +3435\\ \hline \mathbf{254571034}\\ \uparrow \end{array}$$

Review the steps above to convince yourself that they make sense and follow one step to the next. Math is logical; the steps involved flow logically, one to another.

Let's try one more example to make sure we understand the process. It's important to remember that the addition recipe does not change regardless of the actual numbers being added. For example, let's add 999 and 1,000,000. First, write the numbers vertically, and, as the order we add doesn't matter, we'll always write the larger number on top. How do we know which is larger? The number with more digits is larger. Therefore, we write,

$$\begin{array}{r} 1000000\\ +999\\ \hline \end{array}$$

Take a look at this problem. We can work through the algorithm from right to left adding digit by digit. That approach always works. However, do you see a way to get the answer almost immediately? We know that adding zero to a number changes nothing. Above, the first three columns are all $0+9=9$. So, the first three columns of the answer are all 9, there are no carries. The rest of the larger number falls down into the answer to give,

$$\begin{array}{r} 1000000 \\ +999 \\ \hline \mathbf{1000999} \end{array}$$

It always pays to think about the problem before diving in.

However, sometimes we have no option but to plough through it. Consider,

$$\begin{array}{r} 9999999 \\ +99999 \\ \hline \end{array}$$

I recommend you pause here, write this problem on a sheet of paper and work through it before continuing.

For this problem, every column produces a carry. Working right to left then gives us,

$$\begin{array}{r} 11111 \\ 9999999 \\ +99999 \\ \hline 99998 \\ \uparrow \end{array}$$

The ones column is $9+9=18$; the subsequent columns are all $9+9=18$ then $18+1=19$ because of the carry.

At this point, we've used all the digits of the smaller number. To finish the problem, we account for the carry above the next 9 to get $1+9=10$, which itself produces a carry,

$$\begin{array}{r} 111111 \\ 9999999 \\ +99999 \\ \hline 099998 \\ \uparrow \end{array}$$

leading to $1+9=10$ to give us the complete answer,

```
      1 1 1 1 1 1
      9 9 9 9 9 9 9
   +     9 9 9 9 9
   ─────────────────
    1 0 0 9 9 9 9 8
           ↑
```

As before, pause here and convince yourself that the steps above make sense to you.

2.4.1 Section Summary

In this section, we examined adding two numbers with an arbitrary number of digits. As with two two-digit numbers, we write the problem vertically, line up the ones columns, and work right to the left, column by column, moving carries to the next column to the left. This process works to add any two (positive) numbers, regardless of their values. I slipped "positive" into the previous sentence. We'll work with negative numbers in a later chapter. Remember, positive numbers are the numbers bigger than zero and that adding zero to a number doesn't change it.

2.5 Adding Many Numbers

In the 19th century, most accounting was done by hand. Mechanical calculators existed but were relatively rare. This meant learning how to add many numbers together at one time. So let's learn how to add three or more numbers together on the off chance we fall into a wormhole, end up in the 19th century, and decide to take up accounting. Or, more boring but more likely, we want to add the cost of several grocery items together to make sure we have enough money to buy them.

Overall, the process is the same as with two numbers. However, previously the carry was always one as the largest single-digit addition fact is $9 + 9 = 18$ with a carry of 1. When adding three or more numbers, the carry might be more than one. But that's the only difference between adding two numbers and three or more.

For example, let's add 462, 681, and 198. As always, write the problem vertically and line up the ones columns,

```
     4 6 2
     6 8 1
   + 1 9 8
   ─────────
```

Then begin on the right and add $2+1+8$.

This is the first time we've needed to work with an expression like $2+1+8$ where we are asked to add three digits together. This is also our first exposure to what mathematicians call the *order of operations* or *precedence*. These are terms related to the rules for interpreting a mathematical expression, so everyone understands the same thing when they read it. Order of operations is like grammar; it ensures everyone understands math expressions the same way. We'll encounter order of operations repeatedly throughout this book.

For addition, the rule is: add from left to right. We need to be careful here. I'm not talking about the algorithm for adding numbers column by column. Instead, the rule is to add the pair of numbers on the left, then add the next number to the right to that result and so on. For three numbers, this gives us the following sequence of steps where I'm using an arrow ($\rightarrow$) to show the next step in the process,

$$2+1+8 \rightarrow 3+8 \rightarrow 11$$

because $2+1=3$ and $3+8=11$. If we had a fourth number, we'd add it to the 11 that is the sum of the first three.

For the problem at hand, then, the ones column becomes 1 with a carry of one,

$$\begin{array}{r} 1 \\ 462 \\ 681 \\ +\,198 \\ \hline 1 \\ \uparrow \end{array}$$

The tens column is $1+6+8+9$ including the carry from the ones column. This gives us,

$$1+6+8+9 = 7+8+9 = 15+9 = 24$$

where I've slipped in a new use for the equals sign, =. Recall, the equals sign says the thing on the left is the same quantity as the thing on the right. We can extend this use to a series of expressions where = means all expressions are equal. The equals sign is often used this way to indicate that all four of the expressions above, $1+6+8+9$, $7+8+9$, $15+9$, and 24, all have the same value.

2.5. ADDING MANY NUMBERS

The sum of the tens column digits, including the carry, is 24. Here's something new. The carry isn't 1, it's 2. However, we still use it the same way,

$$
\begin{array}{r}
2\,1 \\
4\,6\,2 \\
6\,8\,1 \\
+\,1\,9\,8 \\
\hline
4\,1 \\
\uparrow
\end{array}
$$

To complete the problem, we add the hundreds column including the carry from the tens. That is, $2+4+6+1 = 6+6+1 = 12+1 = 13$,

$$
\begin{array}{r}
2\,1 \\
4\,6\,2 \\
6\,8\,1 \\
+1\,9\,8 \\
\hline
\mathbf{1\,3\,4\,1} \\
\uparrow
\end{array}
$$

Let's consider another example, this time four numbers of different sizes. Here's the problem,

$$
\begin{array}{r}
7\,7 \\
1\,9\,3\,6 \\
5 \\
+2\,8\,4 \\
\hline
\end{array}
$$

As before, please put the book down and try this problem on your own before continuing.

Ready? First, we add the ones column, $7+6+5+4 = 13+5+4 = 18+4 = 22$, so the ones column of the answer is 2 with a carry of 2,

$$
\begin{array}{r}
2 \\
7\,7 \\
1\,9\,3\,6 \\
5 \\
+2\,8\,4 \\
\hline
2 \\
\uparrow
\end{array}
$$

Now add the tens column include the carry, $2+7+3+0+8$. Wait a second, what's that zero doing in there? The third number we're adding is 5, five ones. There are no tens meaning there are zero tens. One way to imagine how this works is to think of zeros in front of the 5,

$$5 = 05 = 005 = 0005$$

There are as many 0s before the 5 as we need for the problem. Here, the largest number we're adding has four digits, so we can imagine 5 is 0005, meaning no thousands, no hundreds, no tens, and five ones. Alternatively, we can simply ignore the empty space in a particular column. In this view, the tens column sum is 2+7+3+8 to account for the carry from the ones column (2), and the digits in those numbers that *do* have a tens column value.

I hope it's clear that $2+7+3+0+8 = 2+7+3+8$ because adding zero changes nothing. Therefore, the tens column of the answer is $2+7+3+8 = 9+3+8 = 12+8 = 20$, meaning 0 with a carry of 2,

```
    2 2
    7 7
  1 9 3 6
        5
+   2 8 4
---------
      0 2
      ↑
```

To complete the sum, we need to add $2+9+2 = 13$ for the hundreds and $1+1 = 2$ for the thousands because of the carry of 1 from the hundreds,

```
  1 2 2
    7 7
  1 9 3 6
        5
+   2 8 4
---------
  2 3 0 2
    ↑
```

Let's do one more problem before leaving this section. Specifically,

2.5. ADDING MANY NUMBERS 47

$$\begin{array}{r} 4000 \\ 500 \\ 70 \\ +6 \\ \hline \end{array}$$

Look carefully at the problem. We saw one similar to it above. However, in this case, every column has only one digit that isn't zero. As adding zero changes nothing, the sum of each column *is* merely the digit that isn't zero,

$$\begin{array}{r} 4000 \\ 500 \\ 70 \\ +6 \\ \hline \mathbf{4576} \end{array}$$

Let's write this problem horizontally,

$$4576 = 4000 + 500 + 70 + 6$$

Writing a number this way is known as *expanded form*. Thinking about a number in expanded form helps us remember what place notation is telling us. In this case, 4576 is a way of writing a number built by grouping 4 sets of one thousand, 5 sets of one hundred, 7 sets of ten, and 6 ones.

We'll discuss it more thoroughly in a later chapter, but the difference between each digit position, going from right to left, is ten. Every time we have ten sets of the previous digit position, we move to the next digit position to the left. Have ten ones? Then we have one set of ten. Have ten sets of ten? That's one hundred. Ten sets of one hundred? You have one thousand, and so on. By adding together sets of thousands, hundreds, tens, and ones, etc., we can specify any number. Keep this thought in the back of your mind. If it's not clear now, don't be concerned, the fog will dissipate as we continue our exploration of arithmetic.

2.5.1 Section Summary

In this section, we completed our introduction to addition by learning how to add an arbitrary collection of numbers with differing numbers of digits. With the algorithm we learned—line up the ones columns, add columns from right to left, use carries—we can add two or more numbers without much trouble. We concluded the section by mentioning the expanded form of a number to help us remember what place notation is telling us.

2.6 Chapter Summary

This chapter was about addition. Specifically, this chapter was about the addition of positive integers. Building on the idea of number from Chapter 1, we discussed addition from two viewpoints. The first used number as quantity to view addition as grouping collections of things together. The second used number as distance on a number line to view addition as motion along the number line, specifically, motion to the right.

To do addition, we first learned how to add two single-digit numbers, which led to Table 2.1 and introduced us to the idea of a carry. We then learned the algorithm to add two two-digit numbers.

After adding two two-digit numbers, we learned how to add two numbers with an arbitrary number of digits and from there how to add three or more numbers regardless of how many digits. The same algorithm works in all cases.

We concluded the chapter by recognizing that writing a number is the same as adding the expanded form of the number. Thus, a number is expressed as a collection of groups where each digit position refers to a set of ten of the digit position immediately to the right.

Let's move on now to the second of our four arithmetic operations, subtraction.

2.7 Terms and Concepts

We introduced the following terms and concepts in this chapter.

Algorithm A recipe. A sequence of steps to perform a task. Addition of two numbers follows a straightforward algorithm of lining up the ones columns and adding digits column by column from right to left, moving carries to the next column as needed.

Carry A carry happens when the sum of two or more digits in a column exceeds nine. The tens column of that sum becomes the carry added to the next column to the left.

Commutative The term mathematicians use to say that the order in which operations happen does not matter. Addition is commutative meaning $2 + 3 = 3 + 2$.

Equals Sign The symbol mathematicians use to indicate that two or more things are the same. If a represents something, and b represents something, then $a = b$ means the two things are the same. For numbers, $2 + 3 = 5$ means $2 + 3$ and 5 are the same quantity.

Expanded Form When a number is written as a sum of digits for each of the columns, $1234 = 1000 + 200 + 30 + 4$. The expanded form of a number makes explicit the meaning of place notation.

Mathematical Equation A mathematical expression using an equals sign to indicate the left and right sides are the same. A mathematical equation is a statement of a fact that two things are the same.

Math Notation The collection of symbols, letters, and numbers that mathematicians use to express mathematical equations and concepts. Math notation forms the alphabet in which the language of mathematics is written.

Operator A mathematical symbol used to indicate an operation like + for addition.

Order of Operations The agreed-upon set of rules used to understand the meaning of mathematical expressions. For addition, the rule is to add from the left to the right, $1 + 2 + 3 = 3 + 3 = 6$.

Plus Sign The mathematical symbol used to indicate addition, +.

Precedence Another name for *Order of Operations*. Sometimes presented as "operator precedence."

Sum Another word for addition, specifically, as a noun, the answer or result of an addition, "the sum of 3 and 4 is 7."

2.8 Exercises

Exercise 1

Draw number lines to represent the following additions.

a) $2+7$
b) $0+1$
c) $7+3$
d) $4+5$

Exercise 2

Solve the following addition problems.

```
    1380          3216          3907          3303
+   3555      +   3147      +    319      +    578
_____      _____      _____      _____

     287          2676           859          3594
+   1535      +   3093      +   1502      +    101
_____      _____      _____      _____

    2012          2333           715          3507
+   1774      +   2154      +    377      +    997
_____      _____      _____      _____

     699           941           343          4012
+   3387      +   3144      +   2185      +    244
_____      _____      _____      _____
```

Think About It

Consider the lengthy addition problem to the right. Our addition algorithm tells us to add the ones column and, if the result is larger than nine, use the tens column of the result as the carry. We did exactly this for all the examples in this chapter. If the sum was, say, 16, we carried the 1. If the sum was 58, we carried the 5, and so on. However, what should we do if the sum of the ones column is not only greater than nine but greater than ninety-nine?

The sum of the ones column on the right is 128, greater than 99. How should we handle this situation? What is the sum of these 15 numbers?

```
      3549
      4278
      8879
      2567
      2399
      4559
      7689
      3898
      6688
      4899
      1049
      3419
      3548
      7549
+      568
_____
```

Chapter 3

Subtraction

Chapter 2 introduced addition as both quantity and motion along the number line. We'll continue that thought in this chapter. Specifically, we'll introduce *subtraction* as "taking away," as removing a smaller quantity from a larger quantity. Then, we'll view subtraction through the lens of motion on the number line.

Next, we'll tackle the subtraction algorithm. Addition introduced the concept of a carry. Subtraction introduces a new concept: borrowing. We'll start slowly by learning how to subtract a smaller number from a larger number using numbers that do not require borrowing. Then, we throw in borrowing. As we'll see, borrowing is only slightly more complex than carrying. Understanding what borrowing means helps us develop a deeper understanding of place notation.

We'll end the chapter by learning how to subtract a larger number from a smaller number. Spoiler: the result is a negative number. Subtraction leading to a negative answer prepares us for Chapter 4's free-for-all approach of adding and subtracting numbers regardless of their sign (positive or negative) and their *magnitude*, that is, how large they are, meaning how far from zero on the number line.

3.1 Subtraction By Taking Away

Addition as merging two (or more) groups was covered in Chapter 2. Here, we view subtraction as taking a smaller group out of a larger group, as taking away.

Consider this example,

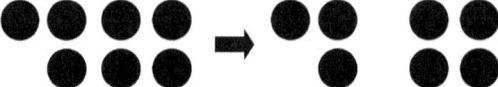

Here, I've taken a larger group of seven and split it into two groups, one of three and the other of four. We saw the reverse of this in Chapter 2 with a group of three and a group of four merged to make the larger group of seven.

The example splits a larger group into two pieces. This is what subtraction does, though, as we'll see, we don't usually write subtraction problems this way. After we subtract a number from another number, we've split the larger number into two groups. The first group has as many items as the number we subtracted. The second group has what's left over, the answer.

Let's present the example above in a slightly different way,

I placed a minus sign between the larger group and the smaller group, and, for good measure, I threw in an equals sign.

We were introduced to the minus sign in Chapter 1 where we used it to indicate a negative number, a number to the left of zero on the number line. Here, we are using the minus sign to indicate subtraction. This is what mathematicians would call an "abuse of notation." Subtraction and marking a negative number are different concepts, but unfortunately, we use the same symbol for both. As with human language, context will have to guide us as to which use is in force.

The equals sign tells us that the example above is an equation that states,

$$7 - 3 = 4$$

In other words, seven minus three equals four.

A quick comment on how we should say $7 - 3 = 4$ is in order. To say the equation out loud, we say "seven minus three equals four," but we don't say we will "minus three from seven." Instead, we use the word *subtract* when we mean the action: "we will subtract three from seven."

Take a second look at the first example above, the one that split seven into two groups, one of three and the other of four. Now look

3.1. SUBTRACTION BY TAKING AWAY 53

at the equation; the three and the four are there. The equation tells us that if we have seven and we want to split it into two groups, one of which has three, then the other group will have four because we know that $3 + 4 = 7$.

The opposite is also true, if we split seven into two groups, one of which is four, then the remaining group will have three, so we can write,

$$7 - 4 = 3$$

The point to remember is that subtraction, viewed as taking away, splits a larger number into two groups and the sum of those two groups is the original number. This helps us understand what subtraction is and, as we'll see later in the chapter, gives us a way to use addition to check if our subtraction answers are correct.

We can write subtraction problems vertically, like we did for addition. Writing vertically makes working out a subtraction problem easier. Therefore, we can write $7 - 3 = 4$ as,

$$\begin{array}{r} 7 \\ -3 \\ \hline 4 \end{array}$$

It may not seem that writing this vertically has helped us here, but it will prove much more useful when we subtract larger numbers later.

We've seen how subtraction is splitting a larger group into two smaller groups. Now let's shift focus and explore subtraction as motion along the number line.

3.1.1 Section Summary

This section discussed subtraction as taking away, as pulling a smaller group from a larger group. We learned that viewing subtraction this way splits a larger group into two pieces and that subtraction is removing a desired number from a larger group and finding what's left behind. We also learned that adding the two smaller groups gives us the original larger group, thereby providing a way to check the subtraction problems we'll work on later in the chapter.

3.2 Subtraction By Moving

Chapter 2 taught us that addition is moving to the right on the number line. We begin at the first number and move to the right the number spaces specified by the second number. Where we end up is the answer, the sum of the first number and second number.

To subtract with the number line, we move *to the left*. We begin at the first number then move a specified distance to arrive at the answer. For example, $7 - 3 = 4$ becomes,

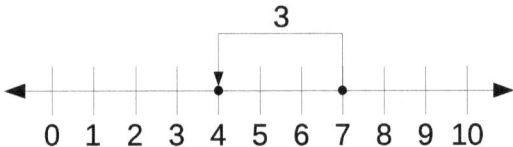

We begin at 7 and move 3 numbers to the left to arrive at 4, the answer.

Why is moving to the left subtraction? Above, we thought of subtraction as "taking away," as separating a larger group into two groups by extracting a smaller group. If we have ten of something, and we take one away, we have nine remaining. For the number line, if we are at ten, meaning a distance of ten from the origin, the zero, and we move one closer to zero, we have taken one away from ten, and we end up at nine: $10 - 1 = 9$. So, moving left some distance from a starting number is subtraction. Also, if we start where we ended, at nine, and move one again to the right, we end up at ten, meaning $9 + 1 = 10$, which is what we expect.

For example, consider this number line,

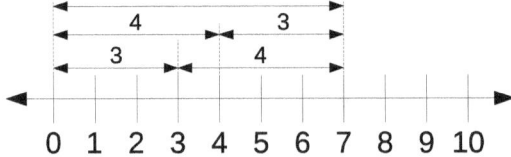

Here, I'm showing three rows of double-sided arrows. The top arrow covers the number line from 0 to 7, so it represents the number 7. The next set of arrows is split. The first covers 0 to 4 and is marked with a 4. The second covers 4 to 7 and is marked with a 3. Together, these arrows also span zero to seven, so $4 + 3 = 7$.

3.2. SUBTRACTION BY MOVING

If we start at 7 and move 3 to the left, we end up at 4, so $7-3=4$. Looking at the arrows, we see that we've split 7 into two pieces, one of size 3 and the other of size 4, just as we did earlier in the chapter. This is the link between the number line and subtraction as taking a smaller group from a larger group.

What do you think the third set of arrows, those closest to the number line, represent? We see that together they also span 0 to 7, so they represent $3+4=7$. Also, if we start at 7 and move 4 to the left, we arrive at 3, so the arrows also represent $7-4=3$.

Now, take a look at the following number line. It's similar to one I showed in Chapter 2. In this case, however, I begin at 10 and subtract first one, then 2, then 3, and so on until subtracting 10,

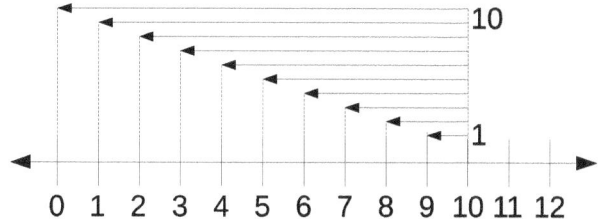

Notice, all arrows face to the left, towards zero.

This diagram presents the subtraction facts for ten. Specifically, it shows us that the following equations are all true,

$$10-0=10$$
$$10-1=9$$
$$10-2=8$$
$$10-3=7$$
$$10-4=6$$
$$10-5=5$$
$$10-6=4$$
$$10-7=3$$
$$10-8=2$$
$$10-9=1$$
$$10-10=0$$

I recommend pausing here for a moment to convince yourself that the equations above are indeed what the number line shows and that you agree with the answers.

Table 3.1: The single-digit subtraction facts

-	0	1	2	3	4	5	6	7	8	9
0	0									
1	1	0								
2	2	1	0							
3	3	2	1	0						
4	4	3	2	1	0					
5	5	4	3	2	1	0				
6	6	5	4	3	2	1	0			
7	7	6	5	4	3	2	1	0		
8	8	7	6	5	4	3	2	1	0	
9	9	8	7	6	5	4	3	2	1	0

Notice two things about the equations and the number line. First, $10 - 10 = 0$, which makes sense from the quantity point of view. If there are ten marbles in a bag and I take ten marbles out, then there are no marbles in the bag, zero marbles. That's the top arrow in the number line above.

Second, the first equation above, $10 - 0 = 10$, tells us that starting at ten and not moving at all keeps us at ten. If we take nothing away from a group, the size of the group doesn't change.

We could make many number lines to illustrate basic subtraction facts, but we'll find it easier to work with a table. Let's examine the first version of it now.

3.2.1 Single-Digit Subtraction Facts

Table 3.1 presents the single-digit subtraction facts, like Table 2.1 does for addition. Let's examine it.

The first thing to notice is that half the table is empty. That's okay. We'll see later in the chapter what goes in the blank spaces, though you might be able to guess, especially if you imagine moving left on the number line past zero.

Let's use the table to subtract five from eight, $8-5$. First, place your finger on the row marked 8. Then, slide along the row until you get to the column labeled 5. The number under your finger is 3, which is the answer, $8-5 = 3$. Use Table 3.1 to find the difference between any two single-digit numbers as long as the first number is larger than the second.

I slipped a new term into the sentence above: *difference*. Sub-

3.2. SUBTRACTION BY MOVING

tracting one number from another is finding the difference between them. So, we might say that the difference between eight and five is three. Therefore, the difference between two numbers is the answer found when subtracting them. Sometimes, the phrase *difference between* is used regardless of the order. The difference between eight and five is three, *and* the difference between five and eight is also three. *Difference between* refers to how far apart two numbers are on the number line, which doesn't change by starting on one number or the other. To think about it another way, the difference between two numbers on the number line is the distance traveled from one to the other.

There is an important item to notice about Table 3.1. For the addition table, Table 2.1, I said it doesn't matter if we start with a row and move to a column or start with a column and move to a row. That's because addition is the same regardless of the order of the two numbers, $3 + 4 = 4 + 3 = 7$.

Subtraction, however, is different: $3 - 4$ is not the same as $4 - 3$. With subtraction, the order matters. Mathematicians would say that subtraction is not commutative; you can't flip the numbers and get the right answer from the table. We'll come back to this observation later in the chapter.

Look again at Table 3.1. Other than the fact that the table is half empty do any patterns jump out to you? What about the diagonal from the upper left to the lower right? Every value in the table is zero. These are the subtraction facts leading to zero, that is, all the cases where the second number is the same as the first. For example, $1 - 1$, $4 - 4$, and $9 - 9$ are all zero. What might the other diagonals, like all the 1's, mean?

3.2.2 Section Summary

In this section, we learned that moving to the left on the number line is subtraction. We then saw some examples, including those for subtracting all single-digit numbers from ten. The idea of moving left to subtract led us to the single-digit subtraction table, at least the first version of it, Table 3.1. With the table, we can find the answer for any single digit subtracted from another single digit, as long as the subtracted digit is smaller.

3.3 Subtraction Without Borrowing

The addition of two numbers, written vertically, moves from right to left, column by column, using single-digit additions with a possible carry to find each column of the answer. The subtraction algorithm works the same way, but we do single-digit subtractions at each column. This section will practice with "nice" numbers, numbers that won't require us to learn about borrowing. We'll save borrowing for the next section.

Let's dive right in and subtract 1723 from 8954,

$$\begin{array}{r} 8954 \\ -1723 \\ \hline \end{array}$$

As before, I'll make the final answer bold as we work from right to left. The algorithm says to line up the ones columns and move from right to left subtracting each column. Therefore beginning with the ones we get,

$$\begin{array}{r} 8954 \\ -1723 \\ \hline 1 \end{array}$$

since $4 - 3 = 1$ according to the number line and Table 3.1.

Moving to the tens column means finding $5 - 2$ which is 3, so we have,

$$\begin{array}{r} 8954 \\ -1723 \\ \hline 31 \end{array}$$

Continuing to the left with the hundreds and then the thousands columns gives us,

$$\begin{array}{r} 8954 \\ -1723 \\ \hline \mathbf{7231} \end{array}$$

This is the final answer.

I claimed above that we can use addition to check if our subtraction result is correct. The problem we just worked is telling us that the difference between 8954 and 7231 is 1723. Thinking of a number line, this means if we start at 7231 and move 1723 numbers to

3.3. SUBTRACTION WITHOUT BORROWING 59

the right, we should end up at 8954. So, to check our answer, we add the result to the number we subtracted. If we're correct, we'll get the number we subtracted from. Addition in this case gives us,

$$\begin{array}{r} 7231 \\ +1723 \\ \hline \mathbf{8954} \end{array}$$

8954 is precisely the number we subtracted 1723 from, so the answer of 7231 is correct. Do spend a bit of time going over this problem again to convince yourself that the steps involved lead to the numbers shown.

Let's work another example, just to make sure everything's working as we expect. How about this one?

$$\begin{array}{r} 3789 \\ -685 \\ \hline \end{array}$$

For maximum effect, pause here and work the problem yourself. Don't forget to refer to Table 3.1 as needed.

To solve this problem, we apply the algorithm moving from right to left, column by column, as before. The first digit of the answer is $9-5=4$. The second digit is $8-8=0$, and the third digit is $7-6=1$. What about the fourth digit, the one for the thousands column? As with addition, the blank is really a zero, there are no thousands in the number 684. The final digit is $3-0=3$. This means the 3 falls down into the answer. So, the solution is,

$$\begin{array}{r} 3789 \\ -685 \\ \hline \mathbf{3104} \end{array}$$

Adding upwards from the answer to the top, and from right to left, we see that $4+5=9$, $0+8=8$, $1+6=7$, and $3+0=3$, to give us 3789, precisely the number we subtracted from, so our answer is correct.

The subtraction algorithm applies regardless of the number of digits in the problem. For example, working out this problem should be, well, not a problem,

$$\begin{array}{r} 987654321 \\ -876543210 \\ \hline \end{array}$$

Copy the problem and work it on a sheet of paper. Or, perhaps you already see what each column subtraction will be?

I recommend you work through a few of the subtraction without borrowing the exercises available as a PDF on the book's website before continuing. You must be comfortable with the subtraction algorithm before proceeding to the next section, which introduces borrowing. (For reference, the answer to the nine-digit subtraction problem above is 111,111,111.)

3.3.1 Section Summary

In this section, we used the single-digit subtraction facts from Table 3.1 along with the subtraction algorithm to find the difference between numbers. The examples were explicitly chosen to produce "nice" subtractions that do not require borrowing, the topic of the next section.

3.4 Subtraction With Borrowing

Consider this subtraction problem,

$$\begin{array}{r} 73 \\ -26 \\ \hline \end{array}$$

The first single-digit subtraction is $3-6$. If we use Table 3.1, beginning with row 3 and sliding over to column 6, we end up at a blank space. The table doesn't help us here. What can we do? Here's where we introduce *borrowing*. As the word implies, we'll borrow a quantity from one place to help us do a single-digit subtraction, the subtraction for the column we are currently working with.

The number we want to subtract 6 from is 73, which, because we know place notation, means seven groups of ten and three ones. Using expanded notation, we can write 73 as,

$$73 = 70 + 3$$

Now, what if I write this,

$$\begin{aligned} 73 &= 70 + 3 \\ &= 60 + 10 + 3 \\ &= 60 + 13 \end{aligned}$$

3.4. SUBTRACTION WITH BORROWING

Here I'm introducing a bit of notation. There are three equations above, all three of which equal 73.

Do you believe me? The first equation says 73 is 70 and 3. The second equation splits the 70 into 60 plus 10. The third equation adds the 10 to the 3 to tell us that $60+13$ also equals 73. In a sense, I borrowed 10 from 70 and gave it to the 3 to make 13.

How does this help us subtract 6 from 73? It helps because I can now change $3-6$ into $13-6$ as long as I change 70 to 60. I'm *borrowing* a ten from the tens column and giving it to the ones column. We might write this vertically as,

$$\begin{array}{r} 6\ 13 \leftarrow \text{after borrowing} \\ \cancel{7}\ \cancel{3} \leftarrow \text{before borrowing} \\ -2\ 6 \\ \hline \end{array}$$

Let's understand what is happening here. First, I borrowed a ten from the tens column and added it to the ones column, $10+3=13$. Then, I removed the ten from the tens column changing the 7 to a 6. Here's where there might be some confusion. If I'm taking *ten*, why change the tens column by only *one*?

Place notation tells us why. The fact that we're working with the tens column is accidental. In place notation, each column to the left of another column counts groups ten times as large. So, to move ten to the current column, we add ten to the current column's value and subtract one from the next column to the left.

For the problem, I show this by striking out the original 7 and 3 replacing them with 6 and 13. The subtraction algorithm is still the same. But, to find the ones column of the answer, we need to find $13-6$, not $3-6$.

Table 3.2 is an extended version of Table 3.1. The rows now go up to 18 to cover subtraction facts up to $18-9$. With Table 3.2, we see that $13-6=7$ by starting at row 13 and moving to column 6.

Our subtraction problem is now,

$$\begin{array}{r} 6\ 13 \\ \cancel{7}\ \cancel{3} \\ -2\ 6 \\ \hline 7 \end{array}$$

And, to complete the solution, we subtract the tens column, $6-2=4$. The final answer is,

Table 3.2: The single-digit subtraction facts extended to 18.

-	0	1	2	3	4	5	6	7	8	9
0	0									
1	1	0								
2	2	1	0							
3	3	2	1	0						
4	4	3	2	1	0					
5	5	4	3	2	1	0				
6	6	5	4	3	2	1	0			
7	7	6	5	4	3	2	1	0		
8	8	7	6	5	4	3	2	1	0	
9	9	8	7	6	5	4	3	2	1	0
10	10	9	8	7	6	5	4	3	2	1
11	11	10	9	8	7	6	5	4	3	2
12	12	11	10	9	8	7	6	5	4	3
13	13	12	11	10	9	8	7	6	5	4
14	14	13	12	11	10	9	8	7	6	5
15	15	14	13	12	11	10	9	8	7	6
16	16	15	14	13	12	11	10	9	8	7
17	17	16	15	14	13	12	11	10	9	8
18	18	17	16	15	14	13	12	11	10	9

3.4. SUBTRACTION WITH BORROWING

$$\begin{array}{r} 6\,13 \\ \not{7}\,\not{3} \\ -2\,6 \\ \hline \mathbf{4\,7} \end{array}$$

To check the answer, find $47 + 26$, which is 73, so we are correct.

The idea of borrowing is a bit strange, but a few more examples should help, along with practice problems involving borrowing. You'll find plenty of those on the book's website.

Give this problem a try before we work it together,

$$\begin{array}{r} 4\,1\,5 \\ -2\,6\,4 \\ \hline \end{array}$$

Remember, borrow when you need to, and the column immediately to the left of the column you are working with is ten times larger, so each borrow adds ten to the current column but subtracts one from the column to the left.

To work this problem, first we need to find $5-4$. This is straightforward, no borrow needed,

$$\begin{array}{r} 4\,1\,5 \\ -2\,6\,4 \\ \hline 1 \end{array}$$

Now we look at the tens column. Here we have to find $1-6$. As 6 is larger than 1, we need to borrow. Where from? The column to the left, the hundreds column. So, we get ten from the hundreds column, add it to the one already in the tens column, and subtract one from the hundreds column,

$$\begin{array}{r} 3\,11 \\ 4\,\not{1}\,5 \\ -2\,6\,4 \\ \hline 1 \end{array}$$

The tens column of the answer is $11-6$ which Table 3.2 tells us is 5,

$$\begin{array}{r} 3\,11 \\ 4\,\not{1}\,5 \\ -2\,6\,4 \\ \hline 5\,1 \end{array}$$

To finish, we need the hundreds column which is now $3-2=1$,

$$
\begin{array}{r}
3\,11 \\
4\,\not{1}\,5 \\
-2\,6\,4 \\
\hline
\mathbf{1\,5\,1}
\end{array}
$$

The check is to add $151+264$, which, if you do it, gives 415 as it should.

3.4.1 Repeated Borrowing

The examples above had us borrow once, meaning one column borrowed from the column to the left, but the subtraction for the column we borrowed from did not itself need to borrow. That isn't always the case.

For example, lets find $1324-578$. The first subtraction we need is $4-8$, which means we need to borrow from the tens as before,

$$
\begin{array}{r}
1\,14 \\
1\,3\,\not{2}\,4 \\
-\ 5\,7\,8 \\
\hline
6
\end{array}
$$

However, the tens column wants us to find $1-7$, for which we need to borrow from the hundreds column. I'll show repeated borrowing like this,

$$
\begin{array}{r}
2\,11 \\
\not{1}\,14 \\
1\,\not{3}\,\not{2}\,4 \\
-\ 5\,7\,8 \\
\hline
4\,6
\end{array}
\quad
\begin{array}{l}
\leftarrow \text{second borrow} \\
\leftarrow \text{first borrow}
\end{array}
$$

The first borrow, the one we needed for the ones column, is marked, then the second borrow for the tens is marked above that. To finish the problem, we need one more borrow to change $2-5$ to $12-5$,

$$
\begin{array}{r}
0\,12 \\
\not{2}\,11 \\
\not{1}\,14 \\
\not{1}\,\not{3}\,\not{2}\,4 \\
-\ 5\,7\,8 \\
\hline
\mathbf{7\,4\,6}
\end{array}
\quad
\begin{array}{l}
\leftarrow \text{third borrow} \\
\leftarrow \text{second borrow} \\
\leftarrow \text{first borrow}
\end{array}
$$

3.4. SUBTRACTION WITH BORROWING

The final digit, the thousands, is $0-0 = 0$, and we don't write leading zeros. Check the answer by adding $746+578$. I recommend walking through the problem again, step by step, to make sure you agree with what each step is doing.

3.4.2 The Trouble With Zero

Consider this problem,

$$\begin{array}{r} 50 \\ -39 \\ \hline \end{array}$$

The first column is $0-9$, which requires a borrow from the tens column. The current value of the ones column is zero, but we add ten to it anyway to give us,

$$\begin{array}{r} 410 \\ \cancel{5}\cancel{0} \\ -39 \\ \hline 1 \end{array}$$

because $10-9 = 1$. To finish the problem, we find the tens column, $4-3 = 1$,

$$\begin{array}{r} 410 \\ \cancel{5}\cancel{0} \\ -39 \\ \hline \mathbf{11} \end{array}$$

Zero wasn't a problem in this case. However, what happens if we need to borrow from a column, but that column is zero? For example,

$$\begin{array}{r} 304 \\ -58 \\ \hline \end{array}$$

The first step in solving the problem is to find $4-8$, which requires a borrow. Okay, the algorithm says to borrow ten from the column to the left and subtract one from that column's value. However, the column to the left is 0; we can't subtract one, so what do we do?

There are several ways to think about what we do next. Perhaps the simplest is to ignore the 0 and look for the first digit to the left that isn't zero. In this case, that's 3. Then, think of the number formed by that digit and all the zeros we skipped to get to it, which would be 30, since we skipped one zero.

We're going to borrow ten from the 30 and subtract one to leave 29. So, we can write this,

$$\begin{array}{r} 2\,9\,14 \\ \cancel{3}\,\cancel{0}\,4 \\ -\ 5\,8 \\ \hline \end{array}$$

Now, we are imagining the top number not as 304, three hundreds and four ones, but as two hundreds, nine tens, and fourteen ones. Therefore, the first digit of the answer is $14-8$, which, according to Table 3.2, is 6, giving us,

$$\begin{array}{r} 2\,9\,14 \\ \cancel{3}\,\cancel{0}\,4 \\ -\ 5\,8 \\ \hline 6 \end{array}$$

We conclude the problem by subtracting the tens column, $9-5=4$, and the hundreds column, $2-0=2$, to arrive at,

$$\begin{array}{r} 2\,9\,14 \\ \cancel{3}\,\cancel{0}\,4 \\ -\ 5\,8 \\ \hline \mathbf{2\,4\,6} \end{array}$$

As the sum of 246 and 58 is 304, we know we have arrived at the correct solution.

To drive home how we intend to handle cases where we need to borrow but the next column to the left is zero, consider this problem,

$$\begin{array}{r} 400001 \\ -198765 \\ \hline \end{array}$$

I recommend you work this one yourself before reading on. Don't be thrown by the number of digits; the algorithm remains the same: work right to left from column one, subtract each single-digit column, borrowing as needed from the column on the left. If you can't borrow, use the "look for the leftmost nonzero digit" trick we used above.

The first column asks for $1-5$, which requires a borrow from the tens column, which is, sadly, zero. Using the trick just mentioned,

3.4. SUBTRACTION WITH BORROWING

we look to the left, column by column, to find the first nonzero column, which is the 4. The number formed by the 4 and all the skipped zeros is 40,000. So, we'll borrow a ten from 40,000, add it to the 1 of column one, and decrement by one to get 39,999. This gives us,

$$\begin{array}{r} 3\,9\,9\,9\,9\,{}^{1}1 \\ 4\,\cancel{0}\,\cancel{0}\,\cancel{0}\,\cancel{0}\,\cancel{1} \\ -1\,9\,8\,7\,6\,5 \\ \hline 6 \end{array}$$

I filled in the first digit of the answer, $11 - 5 = 6$. The remaining columns of the answer are straightforward, no borrowing needed. The answer is,

$$\begin{array}{r} 3\,9\,9\,9\,9\,{}^{1}1 \\ 4\,\cancel{0}\,\cancel{0}\,\cancel{0}\,\cancel{0}\,\cancel{1} \\ -1\,9\,8\,7\,6\,5 \\ \hline \mathbf{2\,0\,1\,2\,3\,6} \end{array}$$

Please spend a bit of time to convince yourself that the answer is correct and that the borrowing with zeros trick makes sense.

What is the borrowing trick doing? When not using the trick, we borrow from the next column to the left, which, because we use place notation based on ten, is ten times bigger than the current column's value. That is, it takes ten of the current column to make up one of the column to the left. When the column to the left is zero, we can imagine borrowing from it anyway by replacing the 0 with the first number to the left of 0 on the number line, -1, negative one. Then, to do the subtraction for the column that is -1, we need to borrow ten from the column to its left, adding ten to the -1, giving us $-1 + 10$.

This is a new concept for us; we haven't explored adding negative numbers yet. Adding ten is moving ten spaces to the right. Moving ten spaces to the right from -1 leaves us at 9,

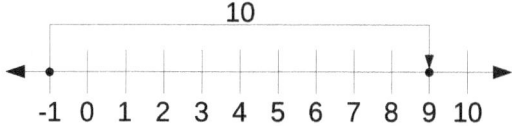

Therefore, $-1 + 10 = 9$, which means after the borrow, the digit becomes a 9. Repeating this for borrows against columns that are 0

leads to the trick above; we get the first nonzero digit, minus one, and replace all the zeros with nines. This is the same as forming the number with the nonzero digit and the skipped zeros, then subtracting one from it, like we did to change 40,000 to 39,999. To me, the trick is simpler to remember, but it is good to know *why* it works. Plus, understanding the trick introduces us to the idea of working with negative numbers, preparing us for the final section of this chapter, to which we now turn.

3.4.3 Section Summary

This section introduced borrowing, the taking of ten from the column to the immediate left of the current column. We worked through the process and what the process means, thereby increasing our understanding of place notation and how it lets us conveniently work with numbers. We then introduced Table 3.2 to extend our subtraction facts to let us look up single-digit subtractions to $18-9=9$, the largest such subtraction we need when borrowing. We concluded the section by learning a trick to handle cases where we need to borrow from a column that is zero and saw that the trick comes from repeated additions of ten to a -1 left after borrowing, thereby introducing our first operation with negative numbers.

3.5 Subtracting a Larger Number From A Smaller Number

What happens if we want to subtract a larger number from a smaller number? We saw this already in the problems above when we needed to borrow. From the subtraction as taking away perspective, we cannot subtract a larger number from a smaller; there aren't enough things to "take away."

However, on the number line, we can subtract a larger number from a smaller. This is because subtraction on the number line is moving from the first number to the left as many spaces as the second number. If we move left of zero, that doesn't matter; we merely arrive at a negative result.

For example, what is $4-7$? To find it with a number line, we start at 4 and move 7 to the left like so,

3.5. SUBTRACTING A LARGER NUMBER FROM A SMALLER NUMBER

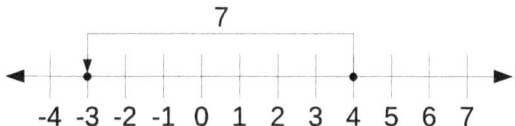

We end at -3, therefore, $4 - 7 = -3$. Similarly, $5 - 9 = -4$, as this number line shows,

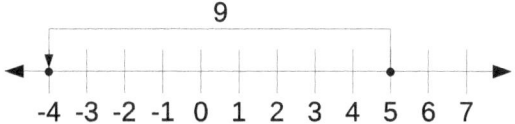

Notice, $4 - 7 = -3$ and $7 - 4 = 3$. Likewise, $5 - 9 = -4$ and $9 - 5 = 4$. These results are not accidents. Here's where we make use of the word *difference* introduced above. The difference between two numbers is the distance between them on the number line. That distance doesn't change if we subtract the smaller from the larger or the larger from the smaller. The only change is that the latter case ends up to the left of zero, meaning the answer is negative.

Excellent! This observation simplifies our lives. If we subtract a larger number from a smaller number, we need only subtract the smaller from the larger and make the answer negative.

Look again at Table 3.1. We now have what we need to fill in the missing entries. They are negative results because we subtract a larger number from a smaller. Table 3.3 presents the complete table.

Let's use Table 3.3 to find $4 - 7$. Begin at row 4 and move over to column 7. Your finger is over -3, precisely what we found with the number line. Similarly, begin at row 5 and move to column 9, your finger is over -4. Now, reverse the problems and find $7 - 4$ and $9 - 5$. You land at 3 and 4, respectively.

There is a symmetry to Table 3.3. Although I don't recommend you do this, if you were to cut the table from the book and fold it along the diagonal from upper left to lower right, you'll find the numbers overlap but are of different signs.

3.5.1 Section Summary

This short section revealed an important fact: subtracting a larger number from a smaller gives the same answer as subtracting the

Table 3.3: The single-digit subtraction facts including negative answers

-	0	1	2	3	4	5	6	7	8	9
0	0	-1	-2	-3	-4	-5	-6	-7	-8	-9
1	1	0	-1	-2	-3	-4	-5	-6	-7	-8
2	2	1	0	-1	-2	-3	-4	-5	-6	-7
3	3	2	1	0	-1	-2	-3	-4	-5	-6
4	4	3	2	1	0	-1	-2	-3	-4	-5
5	5	4	3	2	1	0	-1	-2	-3	-4
6	6	5	4	3	2	1	0	-1	-2	-3
7	7	6	5	4	3	2	1	0	-1	-2
8	8	7	6	5	4	3	2	1	0	-1
9	9	8	7	6	5	4	3	2	1	0

smaller from the larger, but with a negative sign on the answer. Changing the sign of a number is called *negation*. Sometimes, mathematicians refer to the *opposite* of a number, that's the number you get when you change the sign and make a positive number negative or a negative number positive.

3.6 Chapter Summary

We focused on subtraction this chapter. We began with subtraction as taking away. From there, we switched to subtraction as motion along the number line. We then learned the subtraction algorithm and saw that it was very much like the addition algorithm of Chapter 2.

Subtraction inevitably leads to the concept of borrowing, of shifting ten from the column to the left to the current column. We worked with many borrowing examples to handle specific cases, like zeros, that slightly complicate the process. Along the way, we filled in the basic subtraction facts table to arrive at Table 3.2 and Table 3.3.

We concluded the chapter by learning that subtracting a larger number from a smaller number is a meaningful thing to contemplate when thinking of the number line. We then learned that subtracting a larger number from a smaller is the same as subtracting the smaller from the larger and making the answer negative.

3.7 Terms And Concepts

We introduced the following terms and concepts in this chapter.

borrowing The shifting of ten from the column to the left to the current column. Borrowing is to subtraction what carrying is to addition. Borrowing does not change the value of the number, merely how many are in each place when using place notation.

difference The difference between two numbers is the distance between them on the number line. To find this distance, subtract the smaller from the larger.

magnitude The magnitude of a number is its distance from zero on the number line, regardless of whether it is to the left or right of zero. For example, -5 and 5 both have a magnitude of five, but the first is left of zero while the second is to the right of zero.

negation Changing the sign of a number is negating the number. Usually, this means changing a positive number to a negative number by placing the minus sign in front. This is the trick for subtracting a larger number from a smaller one. Negating a negative number produces a positive number.

opposite The number found by negation. The opposite of a positive number is a negative, $4 \to -4$, and the opposite of a negative number is a positive, $-4 \to 4$.

subtraction Subtraction splits a number into two smaller groups if viewed as quantity. If viewed as motion, subtraction moves to the left on the number line from a starting number to a final number. The distance moved is the number subtracted from the first number.

3.8 Exercises

Exercise 1

The following subtraction problems do not require borrowing.

```
  2989        6481        9113        5463
- 2607      - 4211      - 4013      - 5143
  ----        ----        ----        ----

  1348        2739        5343        4572
-  233      -   35      - 4213      - 4071
  ----        ----        ----        ----

  6135        8936        5198        3591
-   33      - 2332      - 2166      - 3550
  ----        ----        ----        ----

  1865        4689        6752        1343
- 1341      - 2058      - 6421      - 1310
  ----        ----        ----        ----
```

Exercise 2

The following subtraction problems require borrowing.

```
  4558        4984        1778        3622
- 4211      - 2528      -  765      -  665
  ----        ----        ----        ----

  2904        2593        4934        1161
- 1800      -  838      - 1899      -  508
  ----        ----        ----        ----

  3113        1606        2847        3468
- 1582      -  990      -  228      -  751
  ----        ----        ----        ----

  3619        3649        4852        3740
- 2552      - 3370      - 2446      - 2295
  ----        ----        ----        ----
```

Exercise 3

The following subtraction problems have multiple zeros.

3.8. EXERCISES

```
   10000            404004
-     499        -   39939
  ───────          ────────


    2001            9001900
-   1002        -    99999
  ───────          ────────
```

Exercise 4

What numbers should go in each square □ to make the equation true? Refer to Table 3.2 as needed.

$$□ - 5 = 7$$
$$18 - □ = 11$$
$$□ - 2 = 15$$
$$7 - □ = 3$$
$$□ - □ = 2$$

For the last equation, find only those pairs of *single-digit* numbers that make the equation true.

Think About It

Earlier in the chapter, we added 10 to -1 to get 9 as the answer, $-1 + 10 = 9$. Addition works if we flip the order, so $10 + -1 = 9$ is also true. However, $10 - 1 = 9$. Therefore,

$$-1 + 10 = 10 + -1 = 10 - 1 = 9$$

Is this an accident, meaning it's only true for 10 and 1 or is it true for all pairs of numbers? If it is true for all numbers, what does that say about subtraction and adding the opposite of a number?

Chapter 4

Mixed-Sign Numbers

The previous chapters approached numbers, addition, and subtraction from two complementary perspectives: number as quantity and number as distance along the number line. In this brief but important chapter, we abandon the quantity perspective and focus instead on the number line and how to use it to add and subtract *mixed-sign* numbers. The phrase "mixed-sign" refers to problems where one number is positive, and the other is negative. We'll also include the case where both numbers are negative.

Our plan of attack is the following. First, we use the number line to investigate the addition of single-digit positive and negative numbers. Second, we do the same for subtraction. When we're done, we'll have a thorough understanding of what addition and subtraction of positive and negative numbers means.

Earlier, we learned algorithms for the addition and subtraction of positive numbers. Understanding what addition and subtraction of mixed-sign numbers means, i.e., what we learn from the number lines, lets us rewrite mixed-sign problems, so we end up working with addition and subtraction as we did in Chapters 2 and 3. No new algorithms are needed.

We end the chapter with a series of example problems involving mixed-sign numbers. In each case, I'll present the problem, work through the reasoning enabling us to rewrite the problem using ordinary addition or subtraction, and, finally, work out the answer.

4.1 Mixing It Up with the Number Line

The number line is a convenient tool for exploring the meaning behind addition and subtraction involving negative numbers. I'll begin with addition using relevant examples and explanations to clarify the process. Then I'll do the same for subtraction. Along the way, we'll learn how to interpret mixed-sign addition and subtraction problems to understand what the problem is telling us, which, in turn, helps us to rewrite the problem so we can use the algorithms of Chapters 2 and 3.

4.1.1 Addition

Chapter 2 taught us that addition is moving to the right along the number line. That remains the case even with negative numbers, with one exception: if the second number is itself negative, the direction changes, and we move *to the left* instead. Let's see some examples and discuss each one to understand why the answer is what it is.

First, let's examine −2 + 4, negative two plus four,

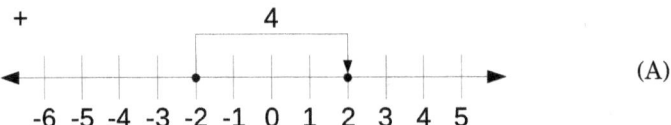

(A)

We start at −2 and move 4 positions to the right to arrive at 2. When the second number is positive, we move to the right, regardless of the sign of the first number. Notice, I put a "+" on the number line to remind ourselves of the operation it represents. I also added a label, "(A)", to refer back to this example later.

Next, consider 5 + −3, five plus negative three. When the second number is negative, you'll sometimes see *parentheses* around the negative number to make the expression easier to read. Therefore, 5 + −3 = 5 + (−3).

Using the number line, we see that 5 + (−3) is,

(B)

4.1. MIXING IT UP WITH THE NUMBER LINE

The number line tells us that $5+(-3) = 2$. Pause a second to let that sink in. Notice, the arrow points to the left, not the right.

We started at 5 and added something, meaning we might expect to arrive at a larger number, somewhere to the right of 5 on the number line. Instead, we moved *left* three to arrive at 2 as the answer. *When adding, and the second number is negative, we move to the left instead of to the right.* The negative sign changes the direction of motion along the number line.

Example (B) above started at a positive number and added a negative number. What happens if we start at a negative number and add a negative number? This number line shows $-3+(-5)$, i.e., negative three plus negative five,

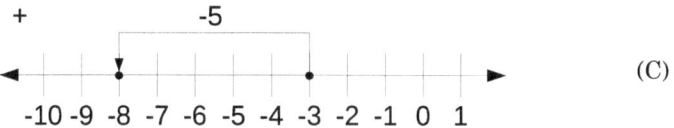
(C)

We end up at -8, meaning $-3+(-5) = -8$. Is this what you expected? Notice, in this case, the arrow points to the left as well. We added a negative number, that we started on a negative number changes nothing, we still move left instead of right because the second number is negative.

Comments

Example (A) showed us that $-2+4 = 2$. In Chapter 2, I said that addition commutes, meaning $3+4 = 4+3$. This holds even with negative numbers. So, we can write,

$$-2+4 = 4+(-2) = 4+-2 = 4-2 = 2$$

I removed the parentheses on $4+-2$ to make the expression look more like $4-2$ because, as it happens, adding a negative is the same as subtraction: $4+-2 = 4-2$. We see this is true in Example (B) above as well, $5+(-3) = 5-3 = 2$.

Example (C) added a -5 to -3 to arrive at -8. We also know that $3+5 = 8$. So, if adding, and both numbers are negative, we can work out the answer by adding the numbers, ignoring the negative signs and making the answer negative when done, i.e., we *negate* the number. We'll use this observation later in the chapter when we add arbitrary mixed-sign numbers.

Before we move on to subtraction, a final comment about how the minus sign is abused, how it's forced to do double-duty: subtraction and negation.

I put parentheses around -3 when adding it to 5 because the notation looks terrible without them: $5 + -3$. We have two operator symbols in a row. Ugly and potentially confusing.

I mentioned in Chapter 1 that some programming languages use a different symbol to mark negative numbers. For example, the programming language APL uses a high minus sign, so in APL, we would enter $5 + {}^-3$ instead of $5 + -3$.

We're not programming computers in APL, so why mention the high minus sign here? Because, when you write problems by hand on paper, you might find it helpful, as I do, to write negative numbers as $^-9$ and not -9. I'll use the high minus sign for negative numbers for the remainder of the chapter, but only when the number after an operator like + or − is negative.

4.1.2 Subtraction

Let's contemplate single-digit subtraction involving negative numbers. We'll use the number line as we did above for addition. The examples will lead us to some helpful observations we can use later in the chapter.

To begin, consider $-3 - 5$, negative three minus five. On the number line, we show the subtraction as,

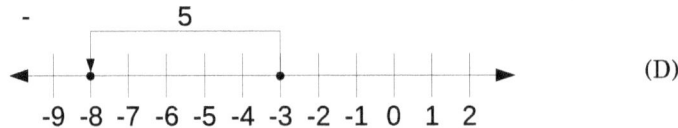

(D)

Notice, we end up at -8, i.e., $-3 - 5 = -8$. Subtraction is moving to the left on the number line. We're subtracting 5 from -3, so we start at -3 and move 5 to the left, just as we did in Chapter 3.

Let's flip the order of the problem to find $5 - (-3)$, or, using the high minus sign, $5 - {}^-3$. The number line for this problem is,

4.1. MIXING IT UP WITH THE NUMBER LINE

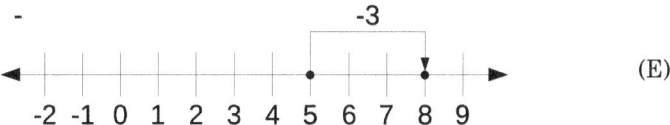

As you may have guessed, we're subtracting a negative number, so we move in the opposite direction, we move to the *right* instead of to the left, $5 - {}^-3 = 8$. Therefore: *when subtracting and the second number is negative, move to the right.*

Finally, here's $-3 - {}^-5$, negative three minus negative five,

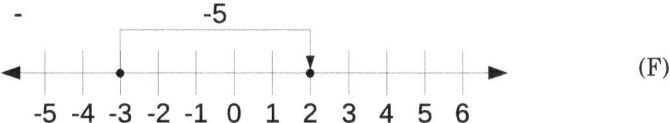

We begin at -3 and subtract -5, which is negative, so instead of moving left, we move right 5 to end up at 2. Therefore, $-3 - {}^-5 = 2$.

Comments

Look again at Examples (C) and (D). The first told us that $-3 + {}^-5 = -8$. The second that $-3 - 5 = -8$. Therefore,

$$-3 - 5 = -3 + {}^-5 = -8$$

From this, we observe that *subtracting a positive number from a negative number is the same as adding two negative numbers.* Whenever I see an expression like $-3 - 5$, I immediately change the minus sign to a plus and use the high-minus to make the expression $-3 + {}^-5$. This reminds me to add the two negative numbers.

Now, look again at Example (E), $5 - {}^-3 = 8$. Subtraction means we should move left on the number line, but the -3 means we change direction and move right instead. But, moving right is addition. Therefore,

$$5 - {}^-3 = 5 + 3 = 8$$

This means subtracting a negative number is really addition. When I see something like $5 - -3$, I change the $--$ into $+$ and add the numbers.

Finally, Example (F) asks us to subtract a negative number from a negative number, $-3 - {}^-5 = 2$. In this case, we can write the following,

$$-3 - {}^-5 = -3 + 5 = 5 + {}^-3 = 5 - 3 = 2$$

Let's take a closer look at this equation to make sure we follow what each part is telling us. First, we have the original problem, $-3 - {}^-5$. To move from that to the second expression, $-3 + 5$, we use the observation we just made above, that $--$ is really addition, $+$.

To move from $-3 + 5$ to $5 + {}^-3$, we use the fact that addition commutes, that it doesn't matter whether we add five to negative three or negative three to five. Finally, from the first observation we made in the addition section, we have that $5 + {}^-3$ is subtraction, $5 - 3 = 2$.

It's worth a bit of time to review the comments here to make sure you understand the reasoning behind them. Ultimately, we'll get to the point where all of the manipulations can be boiled down to a handful of guiding principles.

4.1.3 Section Summary

This section used the number line to solve single-digit addition and subtraction problems involving negative numbers. We can summarize the steps as:

- *Addition* – Begin at the first number. Move to the *right* the distance specified by the second number. If the second number is negative, move *left* instead.

- *Subtraction* – Begin at the first number. Move to the *left* the distance specified by the second number. If the second number is negative, move *right* instead.

We observed that, in some cases, we could change an addition or subtraction problem involving negative numbers into a more convenient form for solving the problem. We'll apply these observations in the next section to help us deal with arbitrary problems involving mixed-sign numbers.

For example, when adding, if one number is positive and the other is negative, then the answer is the same as subtracting the negative number from the positive one, ignoring its sign. E.g.,

$$12345 + {}^-3334 = 12345 - 3334 = 9011$$

and

$$-3334 + 12345 = 12345 - 3334 = 9011$$

We also observed that subtracting a negative number from a positive number is the same as adding ignoring the sign of the negative number,

$$12345 - {}^-3334 = 12345 + 3334 = 15679$$

Finally, we observed that subtracting a positive number from a negative number is the same as adding two negative numbers, which itself is the same as adding the numbers, ignoring their negative signs, and making the answer negative,

$$-3334 - 12345 = -3334 + {}^-12345 = 3334 + 12345 \rightarrow -15679$$

Here, the final expression adds 3334 and 12345, then makes the answer negative to arrive at $-3334 - 12345 = -15679$.

4.2 Addition and Subtraction Free-for-All

The previous section taught us how to work with addition and subtraction problems regardless of the sign of the numbers. The rules and observations are likely somewhat fuzzy. That's to be expected at this point. The cure for fuzzy math thinking is exercise, or, rather, exercises (exercise won't hurt, either). As stated in the introduction: to learn mathematics, we must do mathematics. Therefore, in this section, we put to work everything we've learned up to this point to solve arbitrary addition and subtraction problems. I'll present the problem; then I recommend you put the book down to work it yourself before we work it together.

Problem 1

Find $-271,828 + 141,421$.

This problem asks us to add a positive number to a negative number. We're adding, so we can flip the numbers and write $141,421 + {}^-271,828$. We observed above that adding a negative number is the same as subtracting the positive form of the number. Therefore, to find the answer, we need to find $141,421 - 271,828$. In Chapter 3, we learned that to subtract a larger number from a smaller number we subtract the smaller from the larger and make the answer negative. Therefore, we need to find 271,828 − 141,421 and make the answer negative. Working it out gives us,

$$\begin{array}{r} 271828 \\ -141421 \\ \hline \mathbf{130407} \end{array}$$

Finally, making the answer negative gives us the solution,

$$-271,828 + 141,421 = -130,407$$

This answer makes sense if we think about the number line. We begin at −271,828, which is 271,828 numbers to the left of zero. We then add 141,421. That means we move 141,421 numbers to the right. Moving to the right means we should be closer to zero. We won't go past zero because we're adding a number less than 271,828, we can't move all the way to zero, but we do expect to end up at a negative number that isn't as far from zero as −271,828. We end up at −130,407, which is 130,407 numbers to the left of zero, closer than our starting point of 271,828 numbers to the left of zero.

Problem 2

Find $3,141,592 + {}^-173,205$.

In this case, we're adding a negative number to a positive number. Adding should move right, but the negative second number means move left. Subtraction means move left as well. Therefore, to find the answer, we need to subtract the second number from the first, ignoring its sign: 3,141,592 − 173,205. Another way to see this is to use what we learned in the previous section: +− can be replaced by −. So, subtracting gives us,

$$\begin{array}{r} 2\ 10 \\ \cancel{0}\ 13 \\ \cancel{3}\ 11\ 8\ 12 \\ \cancel{3}\ \cancel{1}\ \cancel{4}\ \cancel{1}\ 5\ 9\ \cancel{2} \\ -\ \ \ 1\ 7\ 3\ 2\ 0\ 5 \\ \hline \mathbf{2\ 9\ 6\ 8\ 3\ 8\ 7} \end{array}$$

We need to borrow multiple times in this problem, so let's walk through the subtraction to make sure understand the steps involved.

4.2. ADDITION AND SUBTRACTION FREE-FOR-ALL

Beginning with the ones column, we need to find $2-5$, which requires a borrow, so we change 9 2 to 8 12 by borrowing ten from the column to the right. We then subtract to find the first two digits of the answer, $12-5=7$ and $8-0=8$.

The next digit of the answer is $5-2=3$. The next digit is $1-3$ requiring another borrow to make 4 1 become 3 11. With that borrow in place, the following digit of the answer is $11-3=8$.

Moving left to the next column means we need $3-7$, requiring yet another borrow. So, we change 1 3 to 0 13 and find the next digit of the answer, $13-7=6$.

One more borrow is needed to do $0-1$. This borrow changes 3 0 to 2 10 thereby allowing us to find the last two digits of the answer: $10-1=9$ and $2-0=2$.

This is the most complex subtraction problem we've yet encountered, but, for all the borrowing, the subtraction algorithm remains the same, we merely need to apply it consistently. To check that our answer is correct, we add,

$$
\begin{array}{r}
1\,1\,11 \\
2\,9\,6\,8\,3\,8\,7 \\
+\ \ 1\,7\,3\,2\,0\,5 \\
\hline
\mathbf{3\,1\,4\,1\,5\,9\,2}
\end{array}
$$

The sum is what we expect, so we have the correct answer. This example warrants a second read, please do to ensure you follow the logic and the steps required to find the answer.

Problem 3

Find $-8,675,309 + {}^-425,066$.

In this case, we begin at a negative number and add another negative number. Adding two negative numbers is the same as adding the positive versions and making the answer negative. Why? Because, addition means move to the right on the number line, but the second number is negative, so we move to the left instead. Moving $425,066$ to the left of $-8,675,309$ leaves us even further from zero, which means we end up at an even more negative number. The number $-8,675,309$ is already $8,675,309$ numbers to the left of zero. Moving another $425,066$ from there means we will be $8,675,309 + 425,066$ to the left of zero. So, we add the positive versions of the numbers and make the answer negative because we are on the left side of zero,

CHAPTER 4. MIXED-SIGN NUMBERS

$$\begin{array}{r} \overset{1\,1\,1\,1}{8675309} \\ +425066 \\ \hline \mathbf{9100375} \end{array}$$

Negating gives us the final answer,
$-8,675,309 + {}^-425,066 = -9,100,375$.

The only difference between adding two negative numbers and adding two positive numbers is the direction we move. For the positive numbers, we move further and further to the right of zero, towards larger positive numbers. For the negative numbers, we move further and further to the left of zero, towards more negative numbers. In both cases, the distance we move from zero is the same, only the direction changes.

Problem 4

Find $-8080 - 6502$.

This problem is of the same form as Problem 3 above. Do you see why? As mentioned previously, I find it helpful to replace minus signs with a plus and make the second number negative: $-8080 - 6502$ becomes $-8080 + {}^-6502$. The problem then becomes one of adding two negatives. However, Problem 3 just showed us how to add two negatives. Therefore, first we add,

$$\begin{array}{r} 8080 \\ +6502 \\ \hline \mathbf{14582} \end{array}$$

Then, we make the answer negative to arrive at $-8080 - 6502 = -14,582$.

Problem 5

Find $161,803 - {}^-231,407$.

Here, we are subtracting a negative number. The "minus a minus" becomes a "plus," meaning $161,803 - {}^-231,407$ becomes $161,803 + 231,407$. So, this problem is nothing more than adding two positive numbers,

4.2. ADDITION AND SUBTRACTION FREE-FOR-ALL

$$\begin{array}{r} 11\\ 161803\\ +231407\\ \hline \mathbf{393210} \end{array}$$

So, $161,803 - {}^-231,407 = 393210$.

Problem 6

Find $-628,319 - {}^-157,079$.

Like Problem 5, we have a "minus a minus" situation, so the problem becomes $-628,319+157,079$. Also, addition lets us flip the numbers if we want to get $157,079 + {}^-628,319$. Adding a negative is subtraction, so we know the answer is the same as the answer to $157,079 - 628,319$, i.e., the $+-$ becomes a $-$.

We're almost there. The subtraction has a larger number subtracted from a smaller number, which we know how to do: subtract the smaller number from the larger, ignoring signs, and make the result negative. Therefore, we first find $628,319 - 157,079$,

$$\begin{array}{r} 5\,12\;\;2\,11\\ 6\;\cancel{2}\;8\;\cancel{3}\;\cancel{1}\;9\\ -\;1\;5\;7\;0\;7\;9\\ \hline 4\;7\;1\;2\;4\;0 \end{array}$$

Then, we take this result and make it negative to arrive at the final answer, $-628,319 - {}^-157,079 = -471,240$.

4.2.1 A Cheat Sheet

When working on mixed-sign addition and subtraction problems, it's best to think through the answer, to think about the number line and how to move along it, and what happens when the second number is negative. That said, we can summarize the steps for solving mixed-sign numbers in a straightforward table that covers all cases for addition and subtraction.

Table 4.1 is such a table. It's a cheat sheet. I strongly encourage you to refer to it *after* you've convinced yourself that you understand the six examples in this section.

The table presents rules for each possible combination of signs of the numbers being added or subtracted. The downside of a cheat sheet is that it lacks context. It doesn't show the reasons, the *why*,

Addition:

pos + pos	Add normally.
pos + neg	Subtract the second from the first.
neg + pos	Subtract the first from the second.
neg + neg	Add ignoring signs. Negate answer.

Subtraction:

pos − pos	Subtract normally.
pos − neg	Add ignoring sign of the second.
neg − pos	Add ignoring signs. Negate answer.
neg − neg	Subtract first from second ignoring signs.

Table 4.1: Mixed-sign addition and subtraction cheat sheet. The first two columns are the signs of the first and second number, "pos" if positive, "neg" if negative.

and it's always best to understand the *why* before resorting to a cheat sheet.

The book's website has many more mixed-sign example problems for you to practice with. Please do. When you are comfortable with these kinds of problems, then you are ready for multiplication, the subject of Chapter 5.

4.2.2 Section Summary

This section presented and solved six mixed-sign addition and subtraction problems. The problems serve as exemplars for the different cases encountered when working with addition and subtraction problems in the wild. The section ended with a cheat sheet, a guide to approaching mixed-sign problems.

4.3 Chapter Summary

In this chapter, we used the number line to explore mixed-sign addition and subtraction. Beginning with single-digit problems, we saw how the number line helps us understand the meaning of addition and subtraction involving negative numbers.

We made several observations about the number line results letting us rewrite mixed-sign problems so we can solve them using the addition and subtraction algorithms we learned earlier in the book.

4.4. TERMS AND CONCEPTS

To test our understanding, we solved six example problems. For each problem, we walked through the logic used to rewrite the problem as an ordinary addition or subtraction. Then, we solved the problem to arrive at the correct answer.

The observations and experience we gained solving the example problems led to Table 4.1, which summarizes the approaches we should take for different mixed-sign problems. As with every such cheat sheet, it is critical to ensure you understand the context and meaning instead of merely applying the rules without understanding.

4.4 Terms And Concepts

We introduced the following terms and concepts in this chapter.

Mixed-sign A mixed-sign problem is one where the signs of the two numbers are not the same. In this chapter, we also included the case where both numbers are negative.

Negate The act of changing the sign of a number to make a negative number positive or a positive number negative.

Parentheses Parentheses set off an expression from other expressions. This chapter used parentheses to separate a negative number from an operator symbol, like −, to make an expression easier to read.

4.5 Exercises

Exercise 1

Solve the following addition problems.

CHAPTER 4. MIXED-SIGN NUMBERS

```
    -455           576           914          -352
 +   210        + -684        +  -56       +  -420
 _____         _____         _____       _____


    -537          -684          -805           673
 +   700        + -528        +  781        +  228
 _____         _____         _____        _____


     681            23          -380          -803
 +   818        +  649        + -314        + -416
 _____         _____         _____        _____


     829          -108           790          -852
 +   756        + -181        +  778        + -706
 _____         _____         _____        _____
```

Exercise 2

Solve the following subtraction problems.

```
     832           448          -362           538
 -   279        -  833        - -300        -  458
 _____         _____         _____         _____


     401           879          -905           507
 -   235        -  455        -   31        -  907
 _____         _____         _____         _____


    -734          -257           685           340
 -   616        - -481        - -572        -  389
 _____         _____         _____         _____


    -283          -529           130           159
 -  -355        - -943        - -746        -  268
 _____         _____         _____         _____
```

Think About It

Is subtraction real in the same sense that addition is real?

Chapter 5

Multiplication

Multiplication, the third of our big four arithmetic operations, is the subject of this chapter. By the chapter's end, we'll know how to multiply two integers, regardless of their size or signs.

We start by thinking of multiplication as repeated addition. From there, we encounter the multiplication table. It's a bit intimidating at first, but don't worry, we'll find plenty of tricks to chop it down to a manageable size.

Once we've conquered the multiplication table, we work up to the general multiplication algorithm in three steps. First, we learn to multiply a number by a single digit. Second, we learn how to multiply two, two-digit numbers. Third, we learn the general multiplication algorithm, a straightforward extension of multiplying two-digit numbers. Finally, we end the chapter by learning how to multiply when negative numbers are involved.

Multiplication is, honestly, rather straightforward. The part that might cause some pain is learning to multiply two single-digit numbers quickly. The best way to do this involves memorization. But, we can make it relatively painless, so, if you've struggled with multiplication before, fear not, you are more than up to the task if you are still with me at this point in the book.

5.1 From Addition to Multiplication

We can think of counting as adding groups of objects where each group contains one item,

$$1 = 1$$
$$2 = 1+1$$
$$3 = 1+1+1$$
$$4 = 1+1+1+1$$
$$5 = 1+1+1+1+1$$

As we learned in Chapter 1, the pattern above matches how the ancients used tally marks.

Now, what happens if the groups we are adding together contain more than one item? Let's add groups of two, not one,

$$2 = 2$$
$$4 = 2+2$$
$$6 = 2+2+2$$
$$8 = 2+2+2+2$$
$$10 = 2+2+2+2+2$$

Repeated additions of the same number is *multiplication*. For example, I can write the above as,

$$2 = 2 \times 1 = 2$$
$$4 = 2 \times 2 = 2+2$$
$$6 = 2 \times 3 = 2+2+2$$
$$8 = 2 \times 4 = 2+2+2+2$$
$$10 = 2 \times 5 = 2+2+2+2+2$$

where I've introduced a new symbol, ×, to represent multiplication. The middle expression uses the new symbol to represent adding that many groups of two together. Adding one group of two is 2×1, adding two groups of two is 2×2, adding three is 2×3, and so on. The first number is how many items are in each group, and the second number is how many groups are added together. Multiplication can be thought of as a shorthand for repeated addition.

Using the definition of multiplication, what is 6×4? The notation 6×4 means "what is the sum of four groups each of six items?" Therefore,

5.1. FROM ADDITION TO MULTIPLICATION

$$6 \times 4 = 6 + 6 + 6 + 6$$
$$= 12 + 6 + 6$$
$$= 18 + 6$$
$$= 24$$

Here, I'm showing the sequence of steps to add six to itself four times. Recall, when adding many numbers, we add the first two, then the next to that sum, and so forth. Therefore, we see that $6 \times 4 = 24$, or, in words, that six times four equals twenty-four. We often use the word "times" to mean multiplication because the second number tells us how many times to add the first number to itself.

One more example before we move on. What is 4×6, i.e., what is four added to itself six times? To find out, we add,

$$4 \times 6 = 4 + 4 + 4 + 4 + 4 + 4$$
$$= 8 + 4 + 4 + 4 + 4$$
$$= 12 + 4 + 4 + 4$$
$$= 16 + 4 + 4$$
$$= 20 + 4$$
$$= 24$$

Therefore, $4 \times 6 = 24$.

Wait a second. Above, we saw that $6 \times 4 = 24$, and now we find that $4 \times 6 = 24$. Have we seen something like this elsewhere? We have, when we learned about addition. We learned in Chapter 2 that addition commutes, that we can flip the order of the numbers, and the answer remains the same. We used this fact extensively in Chapter 4. Now we see that multiplication also commutes, so we can flip the order and still get the same answer. In this case, we see that $4 \times 6 = 6 \times 4 = 24$. This fact will come in handy later in the chapter.

In general, we can write,

$$A \times B = B \times A$$

for any two numbers here represented by A and B. Of course, A and B are letters, not numbers, but we're using them as stand-ins for any two numbers we care to pick. As an aside, thinking this way, i.e., about the form of an expression regardless of the actual numbers involved, is a crucial concept in mathematics beyond arithmetic.

5.1.1 Section Summary

In this section, we learned that multiplication is repeated addition, that some number times a second number is simply adding the first number to itself the second number of times. We also learned that multiplication commutes, that we get the same answer by adding the first number to itself the second number of times, or by adding the second number to itself the first number of times.

5.2 The Multiplication Table

We're about to encounter the multiplication table or times table as it's sometimes called. This table has been a stumbling block for many over the years. It might even be a source of some stress, especially for those who have attempted to master arithmetic in the past. Before discussing the table and tricks to help you learn it, it's best to pause and take a deep breath. The table isn't beyond learning, and if all else fails, and you don't recall a fact, you can certainly work it out by hand as we did above.

Numerous approaches to the multiplication table have been developed over the years. Different people take to different approaches. In this section, I've chosen what I'm calling the minimization approach. First, I'll present the table and then show you tricks you can use to dramatically reduce the number of multiplication facts you need to remember. For some readers, this approach will be ideal; for others, it won't help much at all. That's okay. After discussing the minimization approach, I'll outline other approaches to the table. If tips and tricks aren't your cups of tea, perhaps one of the other approaches will suffice.

In the end, we need to find the answer to any given single-digit multiplication as quickly as possible. How we get the answer doesn't matter. And, as with everything else in mathematics: practice, practice, practice. The book's website contains many multiplication exercises to help you learn the table. The more you practice, the easier it becomes to learn the multiplication facts.

Chapter 2 showed us a table of single-digit addition facts. I encouraged you, primarily via practice problems, to memorize the table. We can make a similar table showing the multiplication facts.

Arithmetic benefits from a bit of rote memorization. The most important things to memorize are the single-digit addition, subtraction, and, now, multiplication facts. Let's examine the table and see

5.2. THE MULTIPLICATION TABLE

×	0	1	2	3	4	5	6	7	8	9
0	0	0	0	0	0	0	0	0	0	0
1	0	1	2	3	4	5	6	7	8	9
2	0	2	4	6	8	10	12	14	16	18
3	0	3	6	9	12	15	18	21	24	27
4	0	4	8	12	16	20	24	28	32	36
5	0	5	10	15	20	25	30	35	40	45
6	0	6	12	18	24	30	36	42	48	54
7	0	7	14	21	28	35	42	49	56	63
8	0	8	16	24	32	40	48	56	64	72
9	0	9	18	27	36	45	54	63	72	81

Table 5.1: The multiplication table

if we can find ways to make it easier to memorize.

Table 5.1 presents the single-digit multiplication table in all its splendor and glory. Let's learn how to use the table before we look for shortcuts. As with the addition facts table, we find any *product*, i.e., the answer for any two numbers multiplied together, by placing our finger on the row with the first number and sliding over the column of the second. So, since we already know that $4 \times 6 = 24$, we can check by starting at row "4" and sliding over to column "6". Doing so means our finger ends up over "24," which is the correct answer. Also, since multiplication commutes, we can start with a column and slide down to a row. Doing so for column "4" down to row "6" also lands at "24."

Okay, now that we know how to use the table, do we really need to memorize all 100 multiplication facts? No, we don't. Let's look for helpful patterns and other shortcuts to make our lives easier.

The first pattern to jump out is what happens when multiplying by zero. The answer is always zero. This is true in general: *multiplying any number by zero gives zero as the answer*. A bit of thought should convince us that this is a reasonable answer. If I add a number to itself zero times, I've added nothing, so I should get zero as the answer. Excellent, this observation removes 19 facts leaving only 81.

The second pattern we see is that *multiplying a number by one gives the number as the answer*, i.e., $3 \times 1 = 3$ and $1 \times 9 = 9$, and so on. Superb! This observation removes another 17 facts from our memorization list. Now we have only 64 facts left; nearly half the table is done for us. Let's see if we can whittle away a bit more.

Here's the remaining multiplication table after removing zero and one,

×	0	1	2	3	4	5	6	7	8	9
0										
1										
2			4	6	8	10	12	14	16	18
3			6	9	12	15	18	21	24	27
4			8	12	16	20	24	28	32	36
5			10	15	20	25	30	35	40	45
6			12	18	24	30	36	42	48	54
7			14	21	28	35	42	49	56	63
8			16	24	32	40	48	56	64	72
9			18	27	36	45	54	63	72	81

Because multiplication commutes, we know that $2 \times 4 = 4 \times 2$, and so on. Can we use this observation here? Yes, if we decide to flip all multiplication problems where the second number is larger than the first, we can clear nearly half of the remaining table. For example, if we see the expression 2×4, we'll mentally flip the problem to become 4×2 because 4 is larger than 2, i.e., $4 > 2$. Doing this leaves the following,

×	0	1	2	3	4	5	6	7	8	9
0										
1										
2			4							
3			6	9						
4			8	12	16					
5			10	15	20	25				
6			12	18	24	30	36			
7			14	21	28	35	42	49		
8			16	24	32	40	48	56	64	
9			18	27	36	45	54	63	72	81

We're down to 36 multiplication facts to memorize. Can we simplify more? Yes.

Look at row 9. Each entry in that row is such that the sum of the digits is always 9. Further, the first digit of the answer is always one less than the number by which 9 is multiplied. So, the answer to any multiplication where the first number is 9 is the second number minus one, followed by whatever we need to add to that value to equal nine.

5.2. THE MULTIPLICATION TABLE

×	0	1	2	3	4	5	6	7	8	9
0										
1										
2			4							
3			6	9						
4			8	12	16					
5			10	15	20	25				
6			12	18	24	30	36			
7			14	21	28	35	42	49		
8			16	24	32	40	48	56	64	
9										

Table 5.2: The bare minimum multiplication facts

For example, we need 9×6. The second number is 6, so the answer starts with 5. What do we need to add to 5 to get 9? We need to add 4. Therefore, we know that $9 \times 6 = 54$. Likewise, $9 \times 2 = 18$ because $2 - 1 = 1$ and we need to add 8 to 1 to get 9. Finally, 9×8 begins with 7 followed by 2 since $7 + 2 = 9$. Therefore, $9 \times 8 = 72$. This observation removes another eight facts from the table. Therefore, at a bare minimum, we need to memorize the 28 single-digit multiplication facts in Table 5.2.

To review, the observations we used to whittle the 100 facts down to 28 are:

- Any number multiplied by zero is zero.
- Any number multiplied by one is the number itself.
- If the first number is smaller than the second, flip the order and multiply the second by the first.
- Nine times any number is the number minus one followed by whatever you need to add to that number to get nine.

How should we learn the remaining facts? A few more observations might help, though, in the end, nothing beats rote repetition.

Look across row 4 of Table 5.2. The three facts in that row are,

$$4 \times 2 = 8$$
$$4 \times 3 = 12$$
$$4 \times 4 = 16$$

Do you notice anything about these three facts? As the second number increases by one, the answer increases by four, i.e., each extra group of four increases the answer by four more than the previous number. In other words, the row counts by fours. This observation holds for all the rows of the multiplication table. The rows are what you get when counting by that row. Thus, for row 6, the increment is six; for row 8, the increment is eight, etc.

This observation is helpful for cases where you might recall that $4 \times 2 = 8$, but you need 4×3 which is currently slipping your mind. You can find 4×3 by adding four to eight because adding one more group of four adds four more. Similarly, if you need 8×6 but only remember that $8 \times 7 = 56$, then you get 8×6 by subtracting 8 from 56 to find that $8 \times 6 = 48$. The counting trick is particularly straightforward for fives. Counting by fives always results in a number ending in either zero or five: 5, 10, 15, 20, etc.

In the end, however, I recommend copying the remaining 28 multiplication facts below by hand, then drill yourself on them in order, over and over, until they come to mind quickly,

2	×	2	=	4	6	×	6	=	36
3	×	2	=	6	7	×	2	=	14
3	×	3	=	9	7	×	3	=	21
4	×	2	=	8	7	×	4	=	28
4	×	3	=	12	7	×	5	=	35
4	×	4	=	16	7	×	6	=	42
5	×	2	=	10	7	×	7	=	49
5	×	3	=	15	8	×	2	=	16
5	×	4	=	20	8	×	3	=	24
5	×	5	=	25	8	×	4	=	32
6	×	2	=	12	8	×	5	=	40
6	×	3	=	18	8	×	6	=	48
6	×	4	=	24	8	×	7	=	56
6	×	5	=	30	8	×	8	=	64

Let's use the facts above and the observations we used to reduce the table, to work through some examples.

Find 4×7

Four is less than seven, so we flip to find 7×4, which is in the set of 28, so we remember that $7 \times 4 = 28$.

5.2. THE MULTIPLICATION TABLE

Find 6×6

This one is also in the table, so we recall that $6 \times 6 = 36$. Note, multiplying a number by itself is called *squaring*, so 36 is equal to 6 *squared*. We'll encounter squaring a number later in Chapter 9.

Find 3×9

Here, we have a choice. Three is smaller than nine, so we might mentally flip the problem to be 9×3, but since it's a number times nine, we can immediately apply the rule for nines. Therefore, the first digit of the answer is $3 - 1 = 2$, and the second is 7 because $2 + 7 = 9$, giving us the answer: $3 \times 9 = 27$.

Find 5×8

For this one, we flip and find 8×5, which is in our set of 28 facts, so we remember that $8 \times 5 = 40$.

Find 5×0

Any number multiplied by zero is zero, so $5 \times 0 = 0$.

Find 7×1

Any number multiplied by one is the number, so $7 \times 1 = 7$.

5.2.1 Options for Learning the Multiplication Table

People have devised many, often novel, approaches to learning the multiplication table. As adults, I think the process I outlined above, namely, a few simple rules and a small set of facts to memorize, is perhaps helpful for many. But, thankfully, we're all different, so one of the options mentioned here might suit you better. The goal is to multiply single digits together quickly. Single-digit multiplication is the heart of the multiplication algorithm we'll learn later in the chapter. The faster we are at it, the easier multiplication problems become. The options here are by no means exhaustive.

Option 1

There is an old story, often claimed to be about the famous 20th-century physicist Niels Bohr. However, it actually appeared in *Pride* magazine in 1958 and happened to American professor Alexander Calandra. Calandra relates that he asked a student to explain how to measure the height of a building using a barometer. The "correct" answer is to use the pressure difference between the top and bottom of the building, which the barometer will show you. However, the student instead offered multiple other answers, all of which were correct but did not include the expected pressure difference reply.

The student's other answers included things like dropping the barometer from the roof and timing how long it takes to hit the ground, tying the barometer to a rope and lowering it from the roof, then measuring the length of the rope, or, my favorite, taking the barometer to the basement office of the building superintendent and offering it as payment for being told the height of the building.

The story's point is that there are often many equally valid ways to achieve the same result. For us, the desired result is to find the product of two single-digit numbers as quickly as possible. The best way to do this is memorization. However, there is also another way, the most obvious way: photocopy Table 5.1 and carry it around with you for those times when you need to multiply two numbers, and you don't have a calculator handy. Use the table even if you have a calculator handy. Why? Because using the table repeatedly will teach you, in time, many of the multiplication facts, eventually getting you to the point where you no longer need the table.

Option 2

Memorization by chanting or singing is a time-honored approach. It's most useful for things like the multiplication table, but should be used with caution lest it becomes a replacement for developing an understanding of what is being memorized.

Many of us learned the alphabet by singing the alphabet song. I'm not immediately aware of an equivalent multiplication song, but I suspect the Internet knows of a few. They may help (caveat emptor). Alternatively, when my children were young and needed to learn their multiplication tables, I had them record themselves reading the table, number by number, in order, as "two times one is two. Two times two is four. Two times three is six." and so on. I then made a compact disc with the recording, one track for each number. Then, often right before bed, the children would play the compact disc and listen as they fell asleep. This approach, anecdotally, seemed to help them remember the table. They used the entire multiplication table, but there is no reason you can't do the

5.2. THE MULTIPLICATION TABLE

same with the voice recorder on your phone and the 28 essential multiplication facts above. For that matter, my smart speaker will recite any multiplication table if asked. Chanting along might help. Or, become annoying as the synthetic voices are good, but not that good. Your mileage may vary.

Option 3

It's been demonstrated that writing class notes by hand versus using a laptop or tablet leads to better scores on tests for questions related to conceptual understanding, see "The Pen Is Mightier Than the Keyboard: Advantages of Longhand Over Laptop Note Taking" by Mueller and Oppenheimer, *Psychological Science*, 2014. For us, conceptual understanding means knowing the multiplication table, including musing over it and learning, by observation and chance recognition, different patterns which we might use, even unconsciously, to recall the multiplication facts.

Applying this knowledge to learning the multiplication table implies that writing the table, either all of Table 5.1 or the essential 28 facts of Table 5.2, will aid in learning. As you write the table on paper, think ahead, meaning if writing "$2 \times 6 = 12$" then think that the next line will be 2×7 and that it will be two larger than the previous, so $2 \times 7 = 14$. Practice each digit until you can write the table quickly and without error.

Option 4

There are all manner of visualizations aiming to help people learn the multiplication table. The visualization I've chosen here is perhaps less helpful for rote memorization but more useful for increasing our understanding, especially for understanding the symmetries buried in the multiplication table.

Figure 5.1 shows the *circle multiplication tables* for the digits one through nine, three per row. This approach was first used, in some form, as part of the Waldorf education process, which emphasizes visual and art-oriented approaches to learning.

To use the tables, begin at "0" then follow the arrow. The arrow points to one times the table's digit. So, for the second table of the top row, the twos, the arrow points to "2" to show that $2 \times 1 = 2$. From that point, the following line moves us to "4." It's the second line traversed, so it's telling us that $2 \times 2 = 4$.

Continuing moves us to "6" to represent $2 \times 3 = 6$ as we've traversed three lines. The next is "8" for $2 \times 4 = 8$, and the following

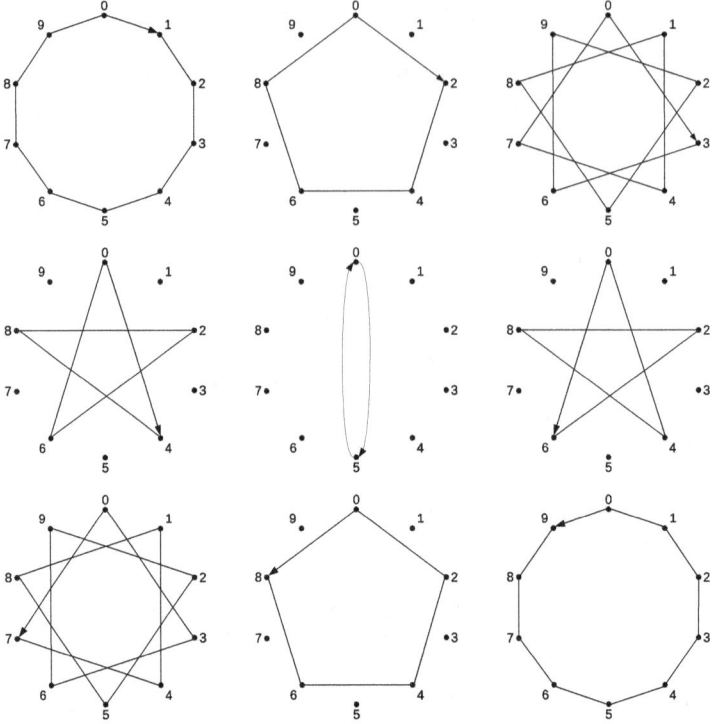

Figure 5.1: Circle multiplication tables. One, two, three (top). Four, five, six (middle). Seven, eight, nine (bottom).

is back to "0" to represent $2 \times 5 = 10$. At this point, the cycle of digits repeats. We've already traversed the circle once, so we continue but add ten to each answer. Therefore, the next position, which has moved across six line segments, represents $2 \times 6 = 10 + 2 = 12$. We can continue cycling forever if we wish, but we need to keep track of how many times we've moved over all the digits and add a ten for each one of them to whatever digit comes next.

What is striking about Figure 5.1 is the symmetry. For example, the patterns for 1 and 9, 3 and 7, and 4 and 6 are identical. Also, the fives table is a compact loop repeatedly jumping from zero to five and back. The symmetries are a consequence of the fact that there are ten points on the circle and that we use ten as the base of our number system. The points on the circles represent the ones value of the rows of the multiplication table.

For numbers greater than five, the arrow points initially to the left instead of the right. For example, for the nines, the first arrow moves to 9 to show $9 \times 1 = 9$. The second segment moves to "8", but adding nine to nine exceeds ten, so we've already cycled, therefore, the second segment shows $9 \times 2 = 10 + 8 = 18$. Of course, we know the rule for nines, but I included nine and one to show the full symmetry of this approach to the multiplication table.

5.2.2 Section Summary

This critical section introduced us to the multiplication table. We need to memorize this table in some fashion to find the product of any two single-digit numbers quickly. To help us along the way, we made several observations about the table, allowing us to reduce it to three rules, multiplying by zero, multiplying by one, the rule for nines, and a core set of 28 multiplication facts.

Acknowledging that not everyone learns in the same way, this section also introduced four optional ways to approach the multiplication table. Thus, all told, we saw five approaches, one of which should, hopefully, help you learn the single-digit multiplication facts.

It's difficult to overemphasize the necessity of mastering the multiplication table. The book's website contains many practice pages. I recommend printing the pages, then timing yourself as you work them to see how quickly and accurately you can solve the problems. As you practice memorizing the table, using whichever option works best for you, you should see your accuracy and speed on the pages improve. Then, when you are comfortable, proceed to the next section.

5.3 Multiplying By Single-Digit Numbers

Let's learn how to multiply a number by a single digit. Knowing how to do this prepares us for the general multiplication algorithm we learn later in the chapter. For example, consider $1,618,034 \times 7$, which we'll write vertically as,

$$\begin{array}{r} 1\ 6\ 1\ 8\ 2\ 3\ 4 \\ \times 7 \\ \hline \end{array}$$

How should we approach this problem? First, think of $1,618,034$ in expanded notation,

```
    1 0 0 0 0 0 0
      6 0 0 0 0 0
        1 0 0 0 0
          8 0 0 0
            2 0 0
              3 0
+                4
    1 6 1 8 2 3 4
```

If we multiply each of the rows above by 7 and add them, we'll get the same answer as multiplying 1,618,234 by 7. This implies we can multiply each digit of 1,618,234 by 7 individually and combine the results much as we did for addition. Indeed, this is precisely what we do: *to multiply a number by a single-digit, multiply each digit from right to left, carrying any tens, then proceed to the next digit adding in any carry from the previous digit.*

When we added, we moved right to the left and added each column. If the result was larger than nine, we carried the tens to the next column to the right, remembering to add any carry when we added that column. For multiplication, we follow the same process, but we *multiply* each digit before adding any carry. Let's solve the example above, and I think you'll see the process is only slightly different from addition.

The first step, then, is to multiply 4×7, which is 28. The 8 goes in the ones column of the answer, and the 2 is carried to the next column,

```
                2
      1 6 1 8 2 3 4
    ×               7
                    8
```

The next digit is $3 \times 7 = 21$, but we need to add the 2 from the previous digit to get $21 + 2 = 23$. The 3 goes in the tens column and the 2 is carried,

```
              2 2
      1 6 1 8 2 3 4
    ×               7
                  3 8
```

The next digit is $2 \times 7 = 14$ with a carry of 2, so we have 16 written as 6 and a carry of 1,

5.3. MULTIPLYING BY SINGLE-DIGIT NUMBERS

$$\begin{array}{r} 1\ 2\ 2 \\ 1\ 6\ 1\ 8\ 2\ 3\ 4 \\ \times 7 \\ \hline 6\ 3\ 8 \end{array}$$

Do you see the pattern? The next digit is $8 \times 7 + 1 = 56 + 1 = 57$,

$$\begin{array}{r} 5\ 1\ 2\ 2 \\ 1\ 6\ 1\ 8\ 2\ 3\ 4 \\ \times 7 \\ \hline 7\ 6\ 3\ 8 \end{array}$$

The next two digits are $1 \times 7 + 5 = 12$ and $6 \times 7 + 1 = 43$. At this point, we're almost finished,

$$\begin{array}{r} 4\ 1\ 5\ 1\ 2\ 2 \\ 1\ 6\ 1\ 8\ 2\ 3\ 4 \\ \times 7 \\ \hline 3\ 2\ 7\ 6\ 3\ 8 \end{array}$$

And, finally, we finish with $1 \times 7 + 4 = 11$,

$$\begin{array}{r} 1\ 4\ 1\ 5\ 1\ 2\ 2 \\ 1\ 6\ 1\ 8\ 2\ 3\ 4 \\ \times 7 \\ \hline \mathbf{1\ 1\ 3\ 2\ 7\ 6\ 3\ 8} \end{array}$$

We have our answer, $1{,}618{,}234 \times 7 = 11{,}327{,}638$. We could check by adding $1{,}618{,}234$ to itself seven times, but I leave that as an exercise for the reader.

Let's work on another problem to make sure we have a handle on the process. We'll find $390{,}435 \times 9$. I suggest you stop here, copy the problem on a sheet of paper, and work through it yourself. Remember, the algorithm is right to left, multiplying digit by digit, and adding any carry from the previous digit.

Ready? Let's do this. I'll jump directly to the answer showing each carry. Walk through the answer yourself and compare, digit by digit, with what you found. If you made an error, no worries, just look back to see what happened. Here's the worked problem,

$$\begin{array}{r} 3\ 8\ 3\ 3\ 4 \\ 3\ 9\ 0\ 4\ 3\ 5 \\ \times 9 \\ \hline \mathbf{3\ 5\ 1\ 3\ 9\ 1\ 5} \end{array}$$

So, $390{,}435 \times 9 = 3{,}513{,}915$. Do practice with the worksheets available on the book's website to convince yourself that you follow the algorithm before proceeding.

5.3.1 Section Summary

This section taught us how to multiply numbers by single digits. The algorithm is quite similar to the one we learned for addition. Specifically,

- Write the problem vertically, lining up the ones columns.
- Work right to left multiplying each digit by the single digit adding in any carry from the previous column to the right.
- Write the ones column in the answer and carry any tens digit.

5.4 Multiplying Two-Digit Numbers

We have one final stop before we conquer the general algorithm for multiplying two numbers. We'll ease into the algorithm by considering a special case: multiplying two, two-digit numbers. Once we understand how to do this and its meaning, we'll be ready for the general algorithm.

I'll present the algorithm for multiplying two-digit numbers via examples. After several, we'll pause and discuss what is happening. I.e., we'll cover the mechanics first, then fill in the theory second.

Find 23×54

First, write the problem vertically, as always,

$$\begin{array}{r} 2\ 3 \\ \times\ 5\ 4 \\ \hline \end{array}$$

How to proceed from here? We learned above how to multiply a number by a single digit. We'll use that knowledge here to multiply 23 by 4 first, then 5. As with all the algorithms we've learned, we work right to left beginning with the ones column. Therefore, we need to find 23×4 which is,

$$\begin{array}{r} 1 \\ 2\ 3 \\ \times\ 5\ 4 \\ \hline 9\ 2 \end{array}$$

So far, so good. Now, we do the same with the 5, but, we write the answer under the 5, not the 4,

5.4. MULTIPLYING TWO-DIGIT NUMBERS

```
      1    ← carry from the 5
      1    ← carry from the 4
      2 3
  ×   5 4
  ───────
      9 2
    1 1 5
```

Let's pause here to explain things. We multiplied 23 by 4; that's the 92 in the problem above. Next, we multiplied 23 by 5 to get the 115 written under the 5, not the 4. There was a carry from multiplying by 4, and another from the 5. I put the carries in separate rows above the problem and labeled them.

To finish the problem, we add the results of the single-digit multiplications,

```
      1    ← carry from the 5
      1    ← carry from the 4
      2 3
  ×   5 4
  ───────
      9 2
  + 1 1 5
  ───────
    1 2 4 2
```

Note, adding $9+5$ produces a carry, which I didn't write, but it's there and leads to $1+1=2$ in the hundreds column. Also, notice that the 2 in the ones column "falls down" into the answer.

Multiplying multi-digit numbers involves the same pattern: multiply individual digits, then add the resulting single-digit answers. Therefore, the form above, multiplication combined with addition, is a standard way of working multiplication problems on paper.

Find 48×61

Again, write the problem vertically,

```
      4 8
  ×   6 1
  ───────
```

Then multiply 48×1,

```
      4 8
  ×   6 1
  ───────
      4 8
```

Notice, as claimed above, multiplying a number by one is just the number, hence the 48 in the partial answer.

Now, multiply 48×6 remembering to write the answer under the 6, not the 1,

$$
\begin{array}{r}
4 \\
4\,8 \\
\times \quad 6\,1 \\
\hline
4\,8 \\
+\,2\,8\,8 \\
\end{array}
$$

Finally, complete the problem by adding,

$$
\begin{array}{r}
4 \\
4\,8 \\
\times \quad 6\,1 \\
\hline
4\,8 \\
+\,2\,8\,8 \\
\hline
\mathbf{2\,9\,2\,8} \\
\end{array}
$$

So, $48 \times 61 = 2928$. And, as above, I'm not showing the carries from the addition step.

Find 99×98

Please try this example yourself before reading through it. Ready? Let's begin by multiplying 99 by 8,

$$
\begin{array}{r}
7 \\
9\,9 \\
\times \quad 9\,8 \\
\hline
7\,9\,2 \\
\end{array}
$$

Now, multiply 99×9 writing the answer below the 9,

$$
\begin{array}{r}
8 \\
7 \\
9\,9 \\
\times \quad 9\,8 \\
\hline
7\,9\,2 \\
+\,8\,9\,1 \\
\end{array}
$$

Then, finish by adding,

5.4. MULTIPLYING TWO-DIGIT NUMBERS

$$\begin{array}{r} 8 \\ 7 \\ 9\ 9 \\ \times\ \ \ 9\ 8 \\ \hline 7\ 9\ 2 \\ +\ 8\ 9\ 1 \\ \hline \mathbf{9\ 7\ 0\ 2} \end{array}$$

We only considered a few examples here, but I hope you see the process. Working through the practice pages on the book's website will help.

5.4.1 Something About Ten

What, exactly, are we doing to multiply two, two-digit numbers? We learned earlier in the chapter how to multiply by a single-digit number. We are doing the same here, once for the ones column, then again for the tens column. However, why are we writing the tens digit answer below the tens digit and not the ones digit? The answer has to do with multiplying by ten or other numbers starting with a one followed by zeros.

Here's a handy observation: *multiplying a number by ten is the same as adding a zero to the number*. This means that $4 \times 10 = 40$ and $1233 \times 10 = 12,330$. We simply add a zero. This is a consequence of using ten as the base of our place notation.

Likewise, multiplying by any number of the form one followed by some number of zeros adds that many zeros: $4 \times 100 = 400$, $13 \times 1000 = 13,000$, and so on.

Now, let's think about 23×54. The number 54 is, in expanded form, $50 + 4$. So, we can think of 23×54 as adding two things, 23×4 and 23×50. We know how to do 23×4. To find 23×50 we observe that $50 = 5 \times 10$, or, in other words, the answer to 23×50 is the same as 23×5 with a zero at the end: $23 \times 5 = 115$, so $23 \times 50 = 1150$.

When we worked this problem above, we wrote,

$$\begin{array}{r} 2\ 3 \\ \times\ \ \ 5\ 4 \\ \hline 9\ 2 \\ +\ 1\ 1\ 5 \\ \end{array}$$

because we wrote the answer to 23×5 below the 5, not the 4.

However, when working with the 5, we aren't really multiplying 23×5, we want 23×50, which is what the 5 means, it's in the tens place. So, if we multiply 23 × 5, we need to add a zero at the end to get 23 × 50 = 1150. By writing the answer to 23 × 5 under the 5, we get 115 but shifted over one place. The 1150 is there, do you see it? It's what we get if we make the blank space a zero, which many people do,

$$\begin{array}{r} 2\ 3 \\ \times\quad 5\ 4 \\ \hline 9\ 2 \\ +\ 1\ 1\ 5\ 0 \\ \hline \end{array}$$

So, when we multiply by the tens place, here a 5, and write the answer below the tens place, we are, in effect, multiplying by 50. Therefore, the two-digit multiplication algorithm is calculating 23 × 4 plus 23 × 50. The addition is the final step.

As an aside, we learned that the answer to a multiplication problem is called the product. When we work multiplication problems as we have here, each of the single-digit answers is known as a *partial product* because it is part of the total product found by adding all the partial products.

5.4.2 Section Summary

Two-digit multiplication was the topic of this section. First, we learned, via examples, how to find the product of two, two-digit numbers. We then learned the meaning behind the steps of the algorithm, why the algorithm is what it is, and what it is doing.

With the multiplication of two, two-digit numbers under our belt, we are now ready for the general multiplication algorithm. Let's see what it entails.

5.5 Multiplying Many-Digit Numbers

Multiplying numbers regardless of how many digits they have is, in the end, a straightforward extension of the algorithm we used to multiply two-digit numbers. I'll illustrate the algorithm via examples; only a few are necessary to clarify the overall process.

Find 359 × 2563

By convention, we write the larger number on top. We can do this because the order we multiply two numbers doesn't matter. Also, by

5.5. MULTIPLYING MANY-DIGIT NUMBERS

placing the larger number on top, we have fewer partial products to find, thereby saving time and reducing our chance of making a mistake. Therefore, let's set up the problem as,

$$
\begin{array}{r}
2\ 5\ 6\ 3 \\
\times 3\ 5\ 9 \\
\hline
\end{array}
$$

We have three digits in the second number (formally called the *multiplier*) and four digits in the first number (the *multiplicand*). Since the multiplier has three digits, we need to find three partial products, one for each digit: 2563×9, 2563×5, and 2563×3. Let's work each one in turn:

$$
\begin{array}{r}
5\ 5\ 2 \\
2\ 5\ 6\ 3 \\
\times 3\ 5\ 9 \\
\hline
2\ 3\ 0\ 6\ 7 \\
\end{array}
$$

Now, 2563×5, with the answer under the 5,

$$
\begin{array}{r}
2\ 3\ 1 \\
5\ 5\ 2 \\
2\ 5\ 6\ 3 \\
\times 3\ 5\ 9 \\
\hline
2\ 3\ 0\ 6\ 7 \\
1\ 2\ 8\ 1\ 5 \\
\end{array}
$$

The last partial product is 2563×3. As you may guess, we write this answer below the 3. The partial product answers are always placed below the digit doing the multiplying. So, we have,

$$
\begin{array}{r}
1\ 1 \\
2\ 3\ 1 \\
5\ 5\ 2 \\
2\ 5\ 6\ 3 \\
\times 3\ 5\ 9 \\
\hline
2\ 3\ 0\ 6\ 7 \\
1\ 2\ 8\ 1\ 5 \\
+\ 7\ 6\ 8\ 9 \\
\hline
\mathbf{9\ 2\ 0\ 1\ 1\ 7} \\
\end{array}
$$

Where I found the 2563×3 partial product, and added all partial products to get the final answer,

$$2563 \times 359 = 920{,}117$$

Find $65,536 \times 64$

First, we set up the problem,

$$\begin{array}{r} 6\,5\,5\,3\,6 \\ \times 6\,4 \\ \hline \end{array}$$

There are two digits in the multiplier, so we need two partial products. The first is,

$$\begin{array}{r} 2\,2\,1\,2 \\ 6\,5\,5\,3\,6 \\ \times 6\,4 \\ \hline 2\,6\,2\,1\,4\,4 \end{array}$$

The second is, including the final addition,

$$\begin{array}{r} 3\,3\,2\,3 \\ 2\,2\,1\,2 \\ 6\,5\,5\,3\,6 \\ \times 6\,4 \\ \hline 2\,6\,2\,1\,4\,4 \\ +\,3\,9\,3\,2\,1\,6 \\ \hline \mathbf{4\,1\,9\,4\,3\,0\,4} \end{array}$$

Therefore, $65,536 \times 64 = 4,194,304$.

Find 1234×4321

Writing the problem with the smaller number as the multiplier gives,

$$\begin{array}{r} 4\,3\,2\,1 \\ \times 1\,2\,3\,4 \\ \hline \end{array}$$

We have four partial products to find. The first two, for 4 and 3, are,

$$\begin{array}{r} 1 \\ 4\,3\,2\,1 \\ \times 1\,2\,3\,4 \\ \hline 1\,7\,2\,8\,4 \\ 1\,2\,9\,6\,3 \end{array}$$

The final two partial products, for 2 and 1, are,

5.5. MULTIPLYING MANY-DIGIT NUMBERS

$$\begin{array}{r} 1 \\ 4\,3\,2\,1 \\ \times \quad 1\,2\,3\,4 \\ \hline 1\,7\,2\,8\,4 \\ 1\,2\,9\,6\,3 \\ 8\,6\,4\,2 \\ +\,4\,3\,2\,1 \\ \hline 5\,3\,3\,2\,1\,1\,4 \end{array}$$

The sum of the partial products gives us the final answer, $4321 \times 1234 = 5,332,114$.

Find 68×1003

Writing the problem vertically with 68 as the multiplier gives,

$$\begin{array}{r} 1\,0\,0\,3 \\ \times \quad 6\,8 \\ \hline \end{array}$$

We need to find two partial products. The first is,

$$\begin{array}{r} 2 \\ 1\,0\,0\,3 \\ \times \quad 6\,8 \\ \hline 8\,0\,2\,4 \end{array}$$

This partial product might throw you a bit at first because of the zeros in 1003. However, nothing about the multiplication algorithm pays any attention to the particular digits involved; we do the same thing each time regardless. So, the first partial product finds $3 \times 8 = 24$, $0 \times 8 + 2 = 2$, $0 \times 8 = 0$, and $1 \times 8 = 8$. Notice, the first multiplication with zero needs to add the 2 from the carry with 3×8, the fact that $0 \times 8 = 0$ changes nothing.

To complete the problem, then,

$$\begin{array}{r} 1 \\ 2 \\ 1\,0\,0\,3 \\ \times \quad 6\,8 \\ \hline 8\,0\,2\,4 \\ +\,6\,0\,1\,8 \\ \hline 6\,8\,2\,0\,4 \end{array}$$

5.5.1 Section Summary

Multiplication involves calculating a series of partial products, one for each digit of the multiplier (the second number), and then adding those partial products together. The algorithm we used above accomplishes exactly this. To be specific, to multiply two numbers by hand,

- Write the problem vertically, lining up the ones columns.

- Work right to left, digit by digit, multiplying the top number (the multiplicand) by the current digit. Record the partial product below, one per line, beginning under the current digit.

- When all partial products have been calculated, add them to arrive at the final answer.

This algorithm works regardless of the order of the numbers, but making the smaller number the multiplier often results in fewer partial products, and who doesn't want to save effort?

That's it. We now know how to multiply two numbers, regardless of their size. Do practice with the worksheets from the book's website. Multiplication is foundational, one of the four arithmetic operations, and one that you are likely to encounter in daily life. For example, your recipe is for two people, but you have six coming to dinner. What do you do? Easy, multiply each ingredient amount by three because $2 \times 3 = 6$. Now, to know you need to multiply by three, and not some other number, involves division, the subject of Chapter 6. But, before we go there, we have one more important multiplication topic to address.

5.6 What About Negative Numbers?

All of the multiplication examples in this chapter use only positive numbers. That's only half the number line. What happens when we multiply with negative numbers? Let's find out.

First, two simple statements or rules to follow using negative numbers. After the rules, we'll discuss them to see why they are what they are. When multiplying two numbers, multiply them, ignoring their signs. Then,

- If *either* of the numbers is negative, the answer is *negative*.

- If *both* of the numbers are negative, the answer is *positive*.

5.6. WHAT ABOUT NEGATIVE NUMBERS?

That's all there is to the mechanics of multiplying numbers when negatives are involved. The rules above are easy to remember; let's look at the *why* behind them.

5.6.1 Why Negative If One Number Is Negative?

To show that the product of a positive number and a negative number is negative, we'll use our old friend, the number line. To get our feet wet, let's see how we might represent the multiplication of two positive numbers using the number line. Specifically, let's represent 2×3. I said at the beginning of this chapter that 2×3 is the same as 2 added to itself 3 times. With a number line, we represent this as,

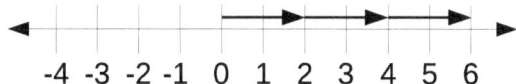

We begin at zero; then we add three groups of two where each "group" is represented by moving a distance of two. Recall, addition using the number line is movement from a starting position. The starting position here is zero because we begin with nothing. Notice that there are three arrows, which all face to the right because addition moves to the right on the number line.

If $2 \times 3 = 6$ is as the above using the number line, what might -2×3 be? First, using the rules above, we know that $-2 \times 3 = -6$ because $2 \times 3 = 6$ and only one of the numbers is negative, so we make the answer negative, -6. How can we represent this using the number line?

One way to think of -2×3 is that we want to move -2 three times in a row. We begin at zero, so moving -2 is moving to the *left*, not the right. Therefore, the number line representation of -2×3 is,

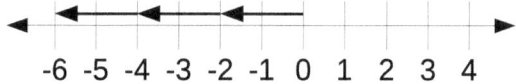

and, as we should, we end up at -6.

I said that the product of a positive and a negative is negative, so what about the other case, $2 \times {}^{-}3$? According to our rules, this should also be -6. How do we represent it on the number line? We represent it by recalling that multiplication is commutative, that we can flip the order. Therefore, $2 \times {}^{-}3 = -3 \times 2$. Written this way, we represent the multiplication in the same way we did for -2×3, only this time we are moving three to the left twice,

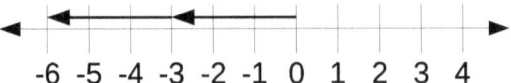

We arrive again at -6, as expected.

5.6.2 Why Positive If Both Numbers Are Negative?

Mathematics is so rich that even when working with something as foundational as arithmetic, it's possible to ask a question that lacks a straightforward answer. "Why is the product of two negative numbers positive?" is one of those questions. The formal proof requires algebra, but, for now, I'll present three intuitive explanations.

The One With The Number Line

We can think of $-2 \times {}^{-}3$ as we thought of adding a negative number in Chapter 4, as something that flips the direction of motion. If -2×3 "wants" to move to the left because of -2, then $-2 \times {}^{-}3$ "wants" to move to the *right* instead. The distance each time we move is still 2, but now to the right. And, we move three times. Therefore, the number line representing $-2 \times {}^{-}3$ is the same as the number line for 2×3,

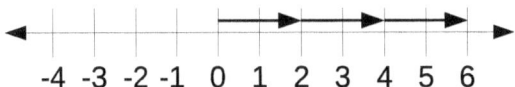

which shows us that $-2 \times {}^{-}3 = 6$, i.e., that the product of two negatives should be a positive number.

The One With The Jogger

Imagine filming someone jogging forward to the right at 2 steps per second. After three seconds, the jogger has moved a distance of 6 steps, i.e., $2 \times 3 = 6$. Now, play the film backward. The jogger appears to be moving backward at 2 steps per second. After three seconds, the jogger is 6 steps to the left. This is $2 \times {}^{-}3 = -6$.

Let's film the jogger again, only this time the jogger, to be quirky, is facing right, as before, but jogging *backward* at 2 steps per second.

5.6. WHAT ABOUT NEGATIVE NUMBERS? 115

After three seconds, the jogger is 6 steps to the left of the starting point. This is $-2 \times 3 = -6$.

Finally, let's play the backward film backward. The jogger was moving backward at 2 steps per second, which we called -2. However, moving the film backward for three seconds we called -3. So, playing the film of the jogger going backward at 2 steps per second for three seconds is $-2 \times {}^-3$. What happens when we watch the film? What we see is the jogger moving *forward* at 2 steps per second for three seconds. Therefore, the jogger ends up at 6, so $-2 \times {}^-3 = 6$.

The One With The Sequence

We know that $-2 \times 3 = -6$ and $-2 \times 2 = -4$ because of the rules above. Also, we know that $-2 \times 3 = {}^-2 + {}^-2 + {}^-2$ and $-2 \times 2 = {}^-2 + {}^-2$. So, -2×2 is one "group" of -2 less than -2×3. This pattern continues. Each time we decrease the second number in the product by one, we are taking one more -2 away from the previous result. Therefore, we can make a table,

-2×3	-6	${}^-2 + {}^-2 + {}^-2$
-2×2	-4	$({}^-2 + {}^-2 + {}^-2) - {}^-2$
-2×1	-2	$({}^-2 + {}^-2) - {}^-2$
-2×0	0	$({}^-2) - {}^-2 = 0$
$-2 \times {}^-1$	2	$(0) - {}^-2 = 0 + 2$
$-2 \times {}^-2$	4	$(0 - {}^-2) - {}^-2 = 0 + 2 + 2$
$-2 \times {}^-3$	6	$(0 - {}^-2 - {}^-2) - {}^-2 = 0 + 2 + 2 + 2$

Some explanation is in order. The first column shows a series of multiplications where the second number decreases by one each time. The second column is the answer. Do you see a pattern? The answer to each multiplication is -2 *less* than the answer to the previous multiplication, just as I stated in the paragraph above. Continuing the pattern moves us from -2×3 to $-2 \times {}^-3$, each time subtracting -2. We end up at $-2 \times {}^-3 = 6$.

The final column on the right shows what is happening. We begin with $-2 \times 3 = {}^-2 + {}^-2 + {}^-2$. Then, -2×2 is written as $-2 \times 3 - {}^-2$ where -2×3 is replaced by ${}^-2 + {}^-2 + {}^-2$ in parentheses. The parentheses make it easier to see that we have $-2 \times 3 - {}^-2$.

Each following row repeats this pattern; the rightmost column is the previous expression minus another -2. When we pass zero, we get $0 - {}^-2$ for $-2 \times {}^-1$. However, we know that subtracting a negative number is addition, i.e., that $--$ becomes $+$, so $-2 \times {}^-1 = 2$, as the expressions on the right of that row indicate. The final two rows follow the pattern until we get to $-2 \times {}^-3 = 6$, as we expect. Therefore, the product of two negative numbers is a positive number.

5.6.3 Section Summary

This section taught us how to multiply positive and negative numbers by introducing two simple rules. First, multiply the numbers ignoring their signs, then,

- If *either* of the numbers is negative, the answer is *negative*.
- If *both* of the numbers are negative, the answer is *positive*.

We then explored three different ways to understand why the product of two negative numbers is a positive number. Specifically, we used the number line, an analogy of a jogger on film, and an argument from the sequence of multiplications with ever-decreasing multipliers (the second number).

5.7 Chapter Summary

In this chapter, we introduced multiplication, perhaps the second most important arithmetic operation after addition. Specifically, we learned that multiplication could be viewed as repeated addition, i.e., adding groups of things together. We then learned the single-digit multiplication table and investigated ways of simplifying it, so we need not memorize all 100 facts. Along the way, we explored several options that might prove helpful when learning the multiplication table.

We then applied the multiplication facts to the task of multiplying a number by a single digit. This introduced a vital component of the general multiplication algorithm, as the general algorithm is repeated multiplications by single digits to generate a set of partial products that are then added together.

We then eased into the general algorithm by first learning how to multiply two two-digit numbers. From there, it was only a tiny step to the general algorithm to multiply two numbers regardless of the number of digits involved.

We concluded the chapter by considering multiplications involving negative numbers. First, we learned that two simple rules decide the sign of a product. Then, we explored three intuitive explanations of why the product of two negative numbers is a positive number.

Addition, subtraction, and now multiplication lay the foundation for our initial foray into the last of the big four arithmetic operations: division, the subject of Chapter 6.

5.8 Terms And Concepts

We introduced the following terms and concepts in this chapter.

multiplier The number multiplying the multiplicand. The second number in the problem.

multiply The repeated addition of a number to itself a given number of times.

multplicand The number being multiplied. The first number in the problem.

partial product The value found by multiplying the multiplicand by a single digit of the multiplier.

product The answer found when two numbers are multiplied. From the Latin *productum* meaning "something produced."

square A number multiplied by itself, e.g. $5 \times 5 = 25$, therefore 25 is 5 squared.

5.9 Exercises

Exercise 1

Solve the following single-digit multiplication problems.

CHAPTER 5. MULTIPLICATION

```
    6          2          9          4
×   5      ×   9      ×   8      ×   4
_____      _____      _____      _____

    4          5          6          7
×   7      ×   2      ×   3      ×   9
_____      _____      _____      _____

    8          3          5          8
×   9      ×   2      ×   7      ×   8
_____      _____      _____      _____

    3          3          7          4
×   5      ×   7      ×   6      ×   8
_____      _____      _____      _____

    7          2          9          9
×   2      ×   4      ×   4      ×   5
_____      _____      _____      _____
```

Exercise 2

Solve the following two-digit multiplication problems.

```
   34         81         17         38
×  79      ×  89      ×  66      ×  36
_____      _____      _____      _____

   57         60         59         24
×  41      ×  41      ×  19      ×  55
_____      _____      _____      _____

   61         76         36         26
×  11      ×  18      ×  14      ×  35
_____      _____      _____      _____

   28         63         31         38
×  39      ×  89      ×  56      ×  95
_____      _____      _____      _____
```

Exercise 3

Solve the following multiplication problems.

5.9. EXERCISES

```
    798          385          180          363
×   555       ×  642       ×  391       ×  312
```

```
    828          829          510          298
×   179       ×  339       ×  945       ×  175
```

```
    312          595          842          421
×   618       ×  686       ×  135       ×  522
```

```
    641          547          834          891
×   742       ×  646       ×  137       ×  816
```

Exercise 4

Is the answer positive or negative?

1. a) $321 \times {}^{-}55$
2. b) -5044×203
3. c) $-4433 \times {}^{-}222$
4. d) -3539×1
5. e) $-3539 \times {}^{-}1$
6. f) -6502×0

Think About It

The ancient Egyptians did not use place notation, yet they could multiply two numbers together. Let's see how by finding 82 × 44. First, let's work out the answer using the technique we learned in this chapter,

```
           8 2
    ×      4 4
           3 2 8
    + 3 2 8
      3 6 0 8
```

Now, let's see how an Egyptian scribe might have worked the problem. The scribe would not have used our digits, but we'll see the process all the same. The basic idea is to make a table with two columns where the numbers on the left are *powers of two*. We'll get to what that means in a later chapter, but each row of the table is the previous number times two. The numbers on the right are the multiplicand where each time it is also multiplied by two, which is nothing more than adding it to itself. So, for 82 × 44, we make a table like so,

$$
\begin{array}{r|l}
1 & 82 \\
2 & 164 \quad = 2 \times 82 \\
4 & 328 \quad = 2 \times 164 \\
8 & 656 \quad = 2 \times 328 \\
16 & 1312 \quad \cdots \\
32 & 2624
\end{array}
$$

I'm explicitly showing how each number in the right column is twice the number above it.

Next, the scribe would look at the numbers in the left column and mark each one that, when added together, makes 44, the multiplier,

$$
\begin{array}{rr|l}
 & 1 & 82 \\
 & 2 & 164 \\
\rightarrow & 4 & 328 \\
\rightarrow & 8 & 656 \\
 & 16 & 1312 \\
\rightarrow & 32 & 2624
\end{array}
$$

because 32 + 8 + 4 = 44. To finish the problem, the scribe would add the numbers in the right column of the marked rows,

```
    3 2 8
    6 5 6
+ 2 6 2 4
---------
  3 6 0 8
```

which is precisely 82 × 44. This process works for any pair of numbers. Why?

Chapter 6

Division

Division, the last of our big four arithmetic operations, is closely related to multiplication. In this chapter, we'll learn what the word "division" means and how to divide arbitrary numbers by a single-digit number. Think of this chapter as preparation for Chapter 7, which explores division in general.

Division answers the question "what must I multiply some number by to get another number as the product?" Consider this equation,

$$6 \times 7 = 42$$

In Chapter 5, we were concerned with the left-hand side of the equation, with finding 6×7. In this chapter, we're concerned with a different form of the equation, with finding the question mark,

$$6 \times ? = 42$$

Whatever the question mark is, it's the number we need to multiply 6 by to get 42 as the answer. Of course, from the equation above, we know that the missing number is 7. So, we can say that 42 divided by 6 is 7, which we write as,

$$42 \div 6 = 7$$

Here, I'm introducing the ÷ symbol as the first of three ways we'll write division problems.

In this chapter, we explore division as repeated subtraction followed by some thoughts on the essence of division. Next, we learn the long division algorithm for dividing an arbitrary number by a single digit. After that, we review the rules for deciding whether or not a single-digit division problem will produce a remainder. We conclude the chapter by learning how to determine the sign of a division problem involving negative numbers.

It might help while reading this chapter to keep the following analogy in mind: division is to multiplication as subtraction is to addition. In each case, the first operation is in some way the opposite of the other.

6.1 Division As Repeated Subtraction

The primary goal of this chapter is to learn the basic form of the long division algorithm. However, to understand what the algorithm is doing, we need to start with the idea of division as repeated subtraction.

In Chapter 5, we learned that multiplication is the repeated addition of groups. Similarly, division can be thought of as the repeated subtraction of groups.

Let's return to the example above, $42 \div 6$. We know the answer is 7 and might see that quickly by looking at the multiplication table. However, what if we didn't see the answer but needed some way to calculate it?

The expression $42 \div 6$ is asking "how many groups of 6, when added together, equal 42?" We can find the answer by repeatedly subtracting 6 and counting how times we subtract before we hit zero. For example,

$$42 - 6 = 36$$
$$36 - 6 = 30$$
$$30 - 6 = 24$$
$$24 - 6 = 18$$
$$18 - 6 = 12$$
$$12 - 6 = 6$$
$$6 - 6 = 0$$

We subtracted 6 seven times, so there are seven groups of 6 in 42, hence, $42 \div 6 = 7$, or,

$$42 = 6 + 6 + 6 + 6 + 6 + 6 + 6$$

6.1. DIVISION AS REPEATED SUBTRACTION

What does division look like on the number line? Let's use a simpler example, one that won't require as many tick marks. Let's visualize $12 \div 3$,

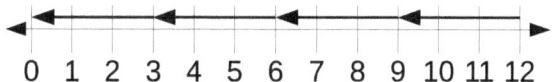

We start at 12, then subtract 3, the first arrow, followed by three more arrows, to arrive at 0. There are 4 arrows, so $12 \div 3 = 4$.

Here's another number line example, $11 \div 3$,

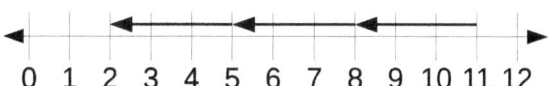

Notice, the arrows stop at 2, not 0. If we added another arrow, if we subtract another 3, we'll go past 0 to -1 because $2 - 3 = -1$. So, we can only subtract three times. We have something left over, meaning 11 cannot be *evenly divided* by 3, meaning 11 is not a *multiple* of 3. Specifically, we have that 11/3 is 3 with a *remainder* of 2.

The paragraph above has three *emphasized* terms. Who knew that 11/3 would cause so much trouble, to say nothing of the new / symbol I slipped in? Let's clarify things.

First, using a slash (/) for division is the second of the three ways we'll write division problems. This is the way you indicate division in algebra, indeed in virtually all of mathematics beyond arithmetic.

Second, we have the new term, *evenly divided*. A number is evenly divided (or, rather, *divisible*) by another if the repeated subtraction of the other ends at zero.

Related to the idea of evenly divisible is that of a *multiple*. One number is a multiple of another if the first number is evenly divisible by the second number. So, 12 is a multiple of 3 because we ended the subtraction process at 0. However, 11 is not a multiple of 3 because we finished at 2, not 0.

Finally, because we finished at 2, we have a remainder of 2; there is a 2 leftover. Therefore, $11 \div 3 = 3$ with a remainder of 2. You'll sometimes see the remainder appended to the answer like so: $11 \div 3 = 3r2$. Read 3r2 as "3 with a remainder of 2."

Let's add a few more new terms since we're in a defining mood. It's best to get division terminology out in the way early on. The parts of a division problem have names,

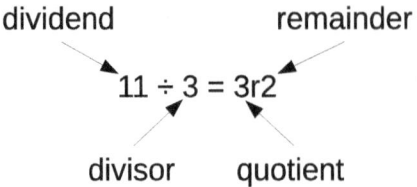

The *dividend* is the number being divided. The *divisor* is the number doing the dividing. The *quotient* is the answer, the number of times the dividend can be split into divisor-sized groups. And, as we already know, the remainder is what is left over; it's the part of the dividend left after we take as many copies of the divisor out of it as possible.

Graphically, 11 ÷ 3 = 3r2 might look like this,

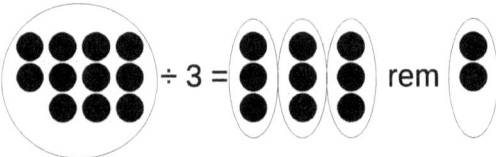

Here, each oval encloses a number, and I'm using "rem" to indicate the remainder, another option you'll encounter from time to time. It's not hard to see that 11 split into groups of 3 gives you three groups with 2 remaining.

6.1.1 A Bit Of Theory (Optional)

For any two integers, which I'll indicate as a and b, the following equation can be made true

$$a = b \times q + r$$

for some other integers, q and r, with $r < b$. When we find q and r to make the equation true, we've found the answer to $a \div b$. As you may guess from the letters I chose, q is the quotient, the answer, and r is the remainder.

Why throw equations with letters in here? Because it helps to think abstractly and not always by remembering an algorithm. Mathematicians use abstract representations of things to show the form that math takes without using a lot of words. Words are tedious and easy to misunderstand. When we learn to read equations,

6.2. THE ESSENCE OF DIVISION

like the above, we'll be able to understand the truths of mathematics without ambiguity. Here, the equation tells us the relationship between quotients and remainders. For example, we've been working with $11 \div 3$. The quotient is 3 and the remainder is 2. We can check if we're right by substituting: $a = 11$, $b = 3$, $q = 3$, and $r = 2$. This gives us,

$$\begin{aligned} 11 &= 3 \times 3 + 2 \\ &= 9 + 2 \\ &= 11 \end{aligned}$$

and it is certainly true that 11 is the same as 11, so we do have the correct answer.

Note, evaluating the equation above uses multiplication before addition. This is an agreed-upon convention, one we haven't encountered before, so it's worth mentioning explicitly.

The equation above only tells us that when we find q and r to make it true for any given a and b that we've solved $a \div b$, it says nothing about *how* to find q and r.

One final comment for those who are familiar with computers and programming languages. One of the operations computers use frequently is called *modulo*, sometimes written as mod or as a character, usually % (percent sign). The modulo operation returns the remainder, the r. So, for our example here, 11 mod 3 = 2 because the remainder after dividing 11 by 3 is 2.

6.1.2 Section Summary

This section introduced division as repeated subtraction. Along the way, we defined a collection of terms related to division: dividend, divisor, quotient, remainder, divisible, multiple, and modulo. It's worth reviewing each term to ensure you understand what it means. We'll be using these terms extensively for the remainder of the chapter (pun intended).

6.2 The Essence of Division

Above, we learned that the answer to a division problem, the quotient, when multiplied by the divisor, is either the dividend or a little less than it, the remainder. If we ignore the remainder for a moment, we see that the quotient times the divisor is the largest multiple of the divisor that is either equal to or less than the dividend.

Therefore, *the goal of division is to find the largest multiple of the divisor that equals or does not exceed the dividend.* How we do this is, in a sense, irrelevant. One possible approach would be to simply guess until we find a multiple equal to the dividend or a multiple closer to the dividend than one divisor.

For simple problems, guessing works. Let's give it a try. What is $74 \div 9$? This means, what multiple of 9 is 74 or within 9 of 74 but not greater?

I'm going to guess 6. $6 \times 9 = 54$. That's 20 away from 74, and 20 is greater than 9, so it isn't 6. If 6 is too small, then let's try something larger; it makes no sense to try anything smaller than 6. How about 9? Well, $9 \times 9 = 81$ and $81 > 74$, so 9 is too large. We need something between 6 and 9. Let's try 8. We know that $8 \times 9 = 72$, and $74 - 72 = 2$, which is certainly less than 9. Therefore, $74 \div 9$ gives us a quotient of 8 with a remainder of 2, or, $74 \div 9 = 8r2$.

Okay, guessing works for small numbers. But, as you may see, it won't be particularly efficient for larger numbers. For example, Uncle Randy passed away at the ripe old age of 107. His little nest egg of $6,283,185$ dollars is to be divided equally among his seven nieces and nephews. Therefore, we must find $6,283,185 \div 7$.

We'll spend a lot of time on trial multiplications if we use guessing, even if we know to bound our guesses like we did above for $74 \div 9$. In fact, a simple computer program implementing the optimum educated guessing approach, known officially as a binary search, takes 21 guesses to find that $6,283,185 \div 7 = 897,597r6$. Most people wouldn't care to do 21 multiplications similar to $897,597 \times 7$ to find the answer to $6,283,185 \div 7$. Then again, if you were one of Uncle Randy's nieces or nephews, you'd have nearly a million reasons to do so if you knew no better way.

How can we do better than guessing? Place notation comes to our aid here. To find the quotient, we need to find each digit of it. The digits of a number tell us how many of that place's value are part of the number. For the example we just solved with guessing on a computer, the first digit of the quotient is 8 but that 8 is in the hundred thousands place, so it isn't 8; it's $800,000$. Similarly, the next digit of the quotient is $90,000$ and so on. If we multiply each digit of the quotient, using its actual place value, by 7 and add the results, we'll get the same value as multiplying $897,597$ by 7, we saw this in Chapter 5.

How does this observation help us divide? We can use the ob-

6.2. THE ESSENCE OF DIVISION

servation to look for the largest multiple of the divisor multiplied by some multiple of different place values. Let's stick with finding $6{,}283{,}185 \div 7$. We'll subtract 7 times each of the place values in the quotient like so,

$$
\begin{aligned}
6{,}283{,}185 - \mathbf{8}00{,}000 \times 7 &= 6{,}283{,}185 - 5{,}600{,}000 &= 683{,}185 \\
683{,}185 - \mathbf{9}0{,}000 \times 7 &= 683{,}185 - 630{,}000 &= 53{,}185 \\
53{,}185 - \mathbf{7}000 \times 7 &= 53{,}185 - 49{,}000 &= 4185 \\
4185 - \mathbf{5}00 \times 7 &= 4185 - 3500 &= 685 \\
685 - \mathbf{9}0 \times 7 &= 685 - 630 &= 55 \\
55 - \mathbf{7} \times 7 &= 55 - 49 &= \mathbf{6}
\end{aligned}
$$

On the left is the expression where the place value of the known quotient is multiplied by 7. I emphasized the quotient digits and the remainder. The middle column shows the actual values subtracted, i.e., $800{,}000 \times 7 = 5{,}600{,}000$. The right column shows what is left over, which becomes the value on the left in the next row. The process continues until all the digits of the quotient have been used. The final value left, 6, is the remainder.

Each row uses the part of the dividend left over because other parts are already accounted for by the quotient digits above that row. For example, The third row works with $53{,}185$ because $6{,}283{,}185 - 5{,}600{,}000 - 630{,}000 = 53{,}185$ with $5{,}600{,}000$ from the 8 in the quotient and $630{,}000$ from the 9 in the ten thousands place. Recall, when you see something like $10 - 3 - 5$, you first do $10 - 3 = 7$ then $7 - 5 = 2$. This same left to right pattern holds for multiplication as well.

This example teaches us that if we know (or can guess) how many of the divisor times each place value fit inside the dividend, or the portion left over after subtracting some previous multiple of the divisor, we have a digit of the quotient. Subtracting that multiple gives us the remaining part of the dividend. Many words, but an example will help. Ultimately, this observation leads to an algorithm that lets us divide any number by a single digit, so bear with me.

Let's try to find $32{,}768 \div 5$. We ask the question: what times $10{,}000$, multiplied again by 5, is as close to $32{,}768$ as possible without exceeding it? Why $10{,}000$? Because there are five digits in $32{,}768$ and we want to know if there are any multiples of $10{,}000$ which, when multiplied by 5, are close to $32{,}768$ without exceeding it. We want to find some digit, call it n, so that $n \times 10{,}000 \times 5$ is as close to $32{,}768$ as possible. We'll pick different n values and see which one works, if any. Remember, n is a placeholder for some number that we're looking to find.

If we pick $n = 1$, meaning we replace n with 1, we get,

$$1 \times 10{,}000 \times 5 = 10{,}000 \times 5 = 50{,}000$$

and $50{,}000 > 32{,}768$, so we know there is no n that fits; the quotient has no digit in the ten thousands place. What about the thousands place? We repeat the process, but now we're looking for a multiple of 1000, something that looks like $n \times 1000 \times 5$. Let's try 6. That gives us,

$$6 \times 1000 \times 5 = 6000 \times 5 = 30{,}000$$

and $30{,}000$ is less than $32{,}768$. We now know that the quotient has a 6 in the thousands place. To find the remaining digits of the quotient, we need to repeat this guessing process for the hundreds digit using what's left over after we subtract 30,000, i.e., now we work with $32{,}768 - 30{,}000 = 2768$.

What n, when multiplied by 100, then 5 (the divisor), is as close to 2768 as possible? Let's guess 6 again. However, $6 \times 100 \times 5 = 600 \times 5 = 3000$, which is too large, so we'll change our guess and try 5: $5 \times 100 \times 5 = 500 \times 5 = 2500$. This fits, so the next digit of the quotient is 5 and we have $2768 - 2500 = 268$ left over.

We need two more digits in the quotient, the tens and the ones. For the tens, let's guess 5 again. This gives us $5 \times 10 \times 5 = 50 \times 5 = 250$, which fits. So the tens digit of the quotient is 5 with $268 - 250 = 18$ remaining.

For the ones, let's guess 3. This gives us $3 \times 1 \times 5 = 3 \times 5 = 15$. Again, this fits, so the final digit of the quotient is 3 with $18 - 15 = 3$ remaining. There are no more digits in the dividend, so the remainder is 3 and the quotient is 6553, therefore, we can write $32{,}768 \div 5 = 6553\text{r}3$.

This process works but is tedious and verbose. However, by altering the form, by introducing a new notation and a compact set of steps, we can wrap this process into a relatively painless algorithm, as the next section will demonstrate.

Therefore, let me explain. No, there is too much. Let me sum up: the steps above uncover the quotient, digit by digit. We move from the largest place value to the smallest, each time guessing to find the proper quotient digit, then subtracting that part (quotient digit × place value × divisor) before proceeding to the next quotient digit until the part of the dividend remaining is less than the divisor.

Returning to our example above, $32{,}768 \div 5$, consider this set of equations, where the digits of the quotient are marked in **bold** text

6.3. DIVIDING BY A SINGLE-DIGIT NUMBER

and the final + 3 is the remainder,

$$32,768 = \mathbf{6} \times 1000 \times 5 + \mathbf{5} \times 100 \times 5 + \mathbf{5} \times 10 \times 5 + \mathbf{3} \times 1 \times 5 + 3$$
$$= 6000 \times 5 + 500 \times 5 + 50 \times 5 + 3 \times 5 + 3$$
$$= 30,000 + 2500 + 250 + 15 + 3$$
$$= 32,768$$

Division uncovers the bold numbers, the digits that go into each place value in the quotient. The reason we work left to right to find the quotient is because after taking the largest amount out of the dividend, here $6 \times 1000 \times 5 = 30,000$, we need only account for what still needs to be added in to close the gap between 30,000 and the dividend, 32,768. We're using the power of place notation to break up the dividend.

6.2.1 Section Summary

This section introduced the goal of division: to find a multiple of the divisor that is as close to the dividend as possible. Anything left over is the remainder. With this goal in mind, we dissected a known division problem to see how the digits of the quotient were extracting multiples of the divisor to account for as much of the dividend as possible.

The dissection led to a process we might use to find the quotient and remainder by moving left to right, digit by digit. We find the largest multiple of the current place value that, when multiplied by the divisor, is as close as possible to the part of the dividend not already accounted for. We applied this process to an example division problem where we produced the quotient, digit by digit, until all the dividend was used, except for the small remainder left over.

6.3 Dividing By a Single-Digit Number

Let's learn how to divide a number by a single digit using several worked examples. It's possible to learn the algorithm without understanding the *why* behind it, but that's boring, and shouldn't satisfy anyone, so please keep the discussion above in mind.

However, before we can work on the algorithm, we need to introduce the last of our three ways of writing a division problem. This way of writing the problem makes it easy for us to work it out by hand. Note, this notation is used primarily in the United States and other parts of North America. Other countries use different notations for division.

Let's walk through the algorithm, step by step, to find $1439 \div 3$. We start by setting up the problem,

$$3\overline{)1439}$$

We write the divisor on the left and the dividend below the tableau (that's what it's called). As we work, we'll put the digits of the quotient above the tableau. The space below the dividend is where we'll work.

To divide, we move from the left to the right. Therefore, we begin with the leftmost digit of the dividend, the 1. We ask the question: "how many times does 3 go into 1?" The question is asking "what do we multiply 3 by to get 1 or as close to 1 as possible without going over 1." The answer is 0 because $0 \times 3 = 0$, and that's less than 1. If we pick 1, we have $1 \times 3 = 3$, which is greater than 1.

When the answer to the question is 0, we move to the next digit to the right. In this case, that's the 4. Now, we ask the question: "how many times does 3 go into 14?" Notice, we keep the 1 as well.

Asking how many 3's go into 14 is more interesting. The answer is 4 because $3 \times 4 = 12$, which is 2 away from 14 and $2 < 3$, so we can't get any closer without going over. How did I see that 3×4 was the closest? By knowing something about the multiplication table and guessing. The division algorithm requires some guessing. So, the first digit of the quotient is a 4. Where does it go? The quotient digits always go above the rightmost digit of the part of the dividend we're working with. In this case, that means the 4 goes over the 4 of "14,"

$$\begin{array}{r} 4 \\ 3\overline{)1439} \\ \underline{12} \\ 2 \end{array}$$

Also, below the "14" we place the 12 that is 3×4. Finally, we draw a short horizontal line and subtract, $14 - 12 = 2$.

The next step requires us to bring down the next digit of the dividend, the 3. This gives us,

6.3. DIVIDING BY A SINGLE-DIGIT NUMBER

$$\begin{array}{r} 4 \\ 3\overline{)1439} \\ \underline{12} \\ 23 \end{array}$$

Now, we ask the question: how many times does 3 go into 23? A bit of guessing should convince you that $3 \times 7 = 21$ and, as that's less than three away from 23, we can't get any closer. Therefore, the next digit of the quotient is 7, and we subtract 21 to get 2,

$$\begin{array}{r} 47 \\ 3\overline{)1439} \\ \underline{12} \\ 23 \\ \underline{21} \\ 29 \end{array}$$

before bringing down the final digit of the dividend, the 9.

To finish we ask: how many 3's in 29? The answer is 9, so we put a 9 in the quotient for the ones digit and subtract $3 \times 9 = 27$ from 29,

$$\begin{array}{r} 479 \\ 3\overline{)1439} \\ \underline{12} \\ 23 \\ \underline{21} \\ 29 \\ \underline{27} \\ 2 \end{array}$$

There are no more digits in the dividend, so we are done. We have the quotient, 479, with 2 left over, which is the remainder: $1439 \div 3 = 479r2$. To check, we need to multiply 479×3, then add the remainder of 2,

$$479 \times 3 + 2 = 1437 + 2$$
$$= 1439$$

This is what we expect. Congratulations! You survived your first *long division* problem. Long division is the name given to the algorithm we just used.

Let's work another single-digit division problem, $7890 \div 6$,

$$6 \overline{)7890}$$

Again, begin on the left and ask: how many times does 6 go into 7? The answer is 1, so the first digit of the quotient is 1. We put $1 \times 6 = 6$ below the 7, subtract, $7 - 6 = 1$, and place the 1 below the line. Finally, we bring down the next digit, the 8,

$$\begin{array}{r} 1 \\ 6{\overline{\smash{\big)}\,7890}} \\ \underline{6} \\ 18 \end{array}$$

To continue, we ask: how many times does 6 go into 18? The answer is 3 because $3 \times 6 = 18$. So, we put 3 in the quotient and 18 below and subtract,

$$\begin{array}{r} 13 \\ 6{\overline{\smash{\big)}\,7890}} \\ \underline{6} \\ 18 \\ \underline{18} \\ 0 \end{array}$$

Notice, the subtraction produced a zero. That's perfectly fine, though perhaps a bit confusing at first. We simply bring down the next digit anyway,

$$\begin{array}{r} 13 \\ 6{\overline{\smash{\big)}\,7890}} \\ \underline{6} \\ 18 \\ \underline{18} \\ 09 \end{array}$$

We don't usually write numbers with leading zeros, like 09, but it's perfectly valid to do so. The zero in front adds nothing other than to be explicit about saying that this number has no tens. Likewise, this number has no ten thousands, either, but we don't usually write $00,009$ because that would be silly; there is no need to specify those digits as they don't contribute to the value of the number. Of course, when the zero is inside the number, then it matters: 409 is $400 + 9$, the middle zero indicating no tens is required.

So, to continue, we ask: how many 6's in 9? The answer is 1 giving us,

6.3. DIVIDING BY A SINGLE-DIGIT NUMBER

$$\begin{array}{r} 131 \\ 6\overline{)7890} \\ \underline{6} \\ 18 \\ \underline{18} \\ 09 \\ \underline{6} \\ 30 \end{array}$$

where, again, I brought down the next digit of the dividend, the 0.

To conclude, we ask: how many times does 6 go into 30? The answer is 5 because $5 \times 6 = 30$. Therefore, we put 5 in the quotient and subtract 30,

$$\begin{array}{r} 1315 \\ 6\overline{)7890} \\ \underline{6} \\ 18 \\ \underline{18} \\ 09 \\ \underline{6} \\ 30 \\ \underline{30} \\ 0 \end{array}$$

We're done. We now know that $7890 \div 6 = 1315$ with no remainder.

Let's work on one more example to help us remember the steps. Division is the most complicated of the arithmetic operations, after all.

Let's find $97{,}804 \div 9$,

$$9\overline{)97804}$$

First, how many 9's go into 9? The answer is 1, so we subtract 1×9 and bring down the next digit, the 7,

$$\begin{array}{r} 1 \\ 9\overline{)97804} \\ \underline{9} \\ 07 \end{array}$$

Now we have something new. How many 9's go into 7? The answer is 0 because 7 is less than 9. Therefore, the next quotient digit is 0.

We could follow the usual steps and subtract 0 from 7 to get 7, then bring down the next dividend digit, the 8, but in this case, we can save a bit of effort and skip the subtraction as it changes nothing. Therefore, we immediately bring down the 8,

$$
\begin{array}{r}
10\\
9{\overline{\smash{\big)}\,97804}}\\
\underline{9}\\
078
\end{array}
$$

Please think carefully about this step to make sure you follow what I did above. To continue, we seek the largest multiple of 9 less than or equal to 78. The answer is $8 \times 9 = 72$, so we have the next digit of the quotient and bring down the 0,

$$
\begin{array}{r}
108\\
9{\overline{\smash{\big)}\,97804}}\\
\underline{9}\\
078\\
\underline{72}\\
60
\end{array}
$$

How many times does 9 go into 60? The answer is 6, i.e., $6 \times 9 = 54$. Therefore,

$$
\begin{array}{r}
1086\\
9{\overline{\smash{\big)}\,97804}}\\
\underline{9}\\
078\\
\underline{72}\\
60\\
\underline{54}\\
64
\end{array}
$$

To finish the problem, we see that $7 \times 9 = 63$, so the final quotient digit is 7,

6.3. DIVIDING BY A SINGLE-DIGIT NUMBER

$$\begin{array}{r}
10867 \\
9{\overline{\smash{)}97804}} \\
\underline{9} \\
078 \\
\underline{72} \\
60 \\
\underline{54} \\
64 \\
\underline{63} \\
1
\end{array}$$

The remainder is 1 giving us the full answer: $97,804 \div 9 = 10,867\text{r}1$.

6.3.1 The Recipe

People learn in different ways. Some prefer to read the recipe; therefore, to divide a number by a single digit, do the following:

1. Work left to right considering each digit of the dividend in turn. If the first digit of the dividend is smaller than the divisor, move to the second and divide the first two digits together.

2. Ask the question: how many X's go into Y? Where X is the divisor and Y the current digit of the dividend. The answer that gets closest to Y is the corresponding digit of the quotient.

3. Multiply the quotient digit and the divisor, then subtract that product from the current dividend digit.

4. Bring down the next digit of the dividend.

5. Repeat from Step 2 until all the digits of the dividend have been examined. The quotient is above the dividend, and any final remainder is below in the work area as what's left over after the final subtraction.

Notice, if we are dividing by 1, we need do nothing, as $1 \times n = n$ for any number you care to put in for n. Therefore, dividing a number by 1 gives you the number as the quotient and no remainder, e.g.,

$$\begin{array}{r}
4576 \\
1{\overline{\smash{)}4576}}
\end{array}$$

I leave working through the division above as an exercise for the reader, should you want to verify my claim.

6.3.2 An Alternative Viewpoint

When we work on long division problems, it's helpful, at least to me, to do as I presented above. However, to increase our understanding of how long division breaks down the dividend until none of it is left, we can show the steps in a slightly different way.

Our first example above was $1439 \div 3$. Let me show you how we worked it on the left, and then another worked version on the right,

$$
\begin{array}{r}
479 \\
3\overline{)1439} \\
\underline{12} \\
23 \\
\underline{21} \\
29 \\
\underline{27} \\
2
\end{array}
\qquad
\begin{array}{r}
479 \\
3\overline{)1439} \\
\underline{1200} \\
239 \\
\underline{210} \\
29 \\
\underline{27} \\
2
\end{array}
$$

The version on the right shows the actual values we subtract at each step. For example, when working on the problem, we used $14 - 12 = 2$, but on the right, we see that it was really $1439 - 1200 = 239$. Likewise, for the next step, it wasn't $23 - 21 = 2$ but $239 - 210 = 29$.

The viewpoint on the right mirrors the rows of subtractions I presented earlier in the chapter. For this problem, we can write,

$$
\begin{aligned}
1439 - 400 \times 3 &= 1439 - 1200 = 239 \\
239 - 70 \times 3 &= 239 - 210 = 29 \\
29 - 9 \times 3 &= 29 - 27 = 2
\end{aligned}
$$

Far from being mysterious, then, the long division algorithm is merely a shorthand way of doing what we reasoned through initially. Some people might find it helpful to work long division as on the right above, by putting the actual product in each row and subtracting across. You still need to work from left to right to get each digit of the quotient, so, in the end, it doesn't save time, but the notation might clarify things a bit. The worksheets on the book's website use the form on the right when presenting solutions to long division problems.

6.3.3 Danger, Will Robinson!

We can now divide any number by any single digit. Or, can we? There are two special cases we need to discuss. The first is dividing any number, other than zero, by zero. The second is dividing zero by zero. Let's examine each in turn.

6.3. DIVIDING BY A SINGLE-DIGIT NUMBER 137

Dividing By Zero Is Not A Thing

I want to find $359 \div 0$. To use the subtraction approach, I need to count how many times I can subtract 0 from 359. However, $359 - 0 = 359$. No matter how many zeros I subtract, 359 never gets any smaller. There is no meaning to the idea of subtracting zero repeatedly until a number is reduced to zero.

What if I persist and try to use the long division algorithm for single digits? Let's set up the problem,

$$0\overline{)359}$$

To begin, I ask: how many times does 0 go into 3? Again, there is no logical answer, I can add zero to itself until the cows come home, grow old, die, rot, and turn to dust, I'll still only ever have zero (and no cows). Dividing a number by zero is, to use the mathematician's word, *undefined*. There is no logic to the operation; it makes no sense.

To be pedantic, you cannot divide by zero when discussing arithmetic, as we are here in this book. However, advanced forms of mathematics admit to a notion akin to division by zero. For us, however, dividing by zero is not a thing.

Zero Divided By Zero Is Indeterminate

We started this chapter with an equation that looked like this,

$$6 \times a = 42$$

where I'm now using the letter a in place of the question mark I used then. The equation is asking: what times 6 equals 42? I said that a is what we get when we divide 42 by 6: $a = 42/6$, to use the slash notation. We learned that 7 is the answer, which means that $6 \times 7 = 42$.

Now, consider this equation,

$$0 \times a = 0$$

It's asking: what times 0 equals 0? Or, in other words, what is $a = 0/0$? What number times zero equals zero? All of them. Pick any number you like; if you multiply it by zero, you get zero. Therefore, any number you pick for a fits in the equation $a = 0/0$. We can't select just one specific a. Mathematicians say that such expressions are *indeterminate*; they can't be pinned down.

Another way to see that 0/0 is indeterminate is to use a bit of algebra. Algebra says that an expression like $a = 0/0$ can be rewritten as $0 \times a = 0$. Specifically, it's what we get by multiplying both sides of the equal sign by 0. However, we just realized that any a satisfies $0 \times a = 0$. So, we can't pick just one a; hence, 0/0 is indeterminate.

This same algebra argument also demonstrates that dividing by zero is meaningless. Let's find $b = 1/0$. Algebra says we can write the expression as $0 \times b = 1$. What times zero equals one? Nothing. There is no solution, no b. Hence, dividing by zero makes no sense.

The moral is clear: *never divide by zero*.

6.3.4 Section Summary

In this section, we learned the long division algorithm for single-digit divisors. We worked several examples together to learn the algorithm by doing, then reviewed a recipe version. We ended the section by learning why we cannot divide by zero.

This section is critical for our understanding of long division. Please review it carefully before proceeding to the next chapter, where we expand the algorithm to include multidigit divisors. As always, the book's website holds a wealth of practice problems for you.

6.4 Will It Divide?

When we say some number, call it x, evenly divides another, call it y, we mean that $y \div x$ has no remainder. Remember, x and y are stand-ins for any numbers we choose. To be more specific, let's restrict x to be a single digit, some number from 2 to 9. When mathematicians say that x *divides* y, they mean that there is no remainder to $y \div x$, i.e., that y is a multiple of x.

When we worked $1439 \div 3$, we found a remainder of 2. Therefore, 3 does not divide 1439. Is there a way to know ahead of time, without doing the division problem, if 3 will divide 1439? It turns out that there is. It's possible to tell without dividing if any single-digit division problem will have a remainder or not. There are rules we can use for each digit from 2 through 9. Let's learn them now. I'll even throw in an extra rule for 10 as a bonus.

Rule of 2

If a number is even, it's divisible by 2. Fine, but what's an *even number*? The definition is a bit circular because an even number is a number that's divisible by 2. To be specific, even numbers end in 0, 2, 4, 6, or 8, all multiples of 2. If a number ends in 1, 3, 5, 7, or 9, it isn't an even number; it's an *odd number*.

So, does $2{,}434{,}006{,}905 \div 2$ have a remainder? Yes, because $2{,}434{,}006{,}905$ ends in 5 and is, therefore, an odd number. Moreover, since remainders must be less than the divisor, and the divisor is 2, the only possible remainders when dividing by 2 are 0 if the number is even, and 1 if the number is odd. Therefore, in this case, the remainder is 1.

Rule of 3

For any number, add the digits together. If that number is divisible by 3, then the original number is also divisible by 3. You can repeat this process until you get a number that you recognize as divisible by 3 or not.

For example, is $39{,}560$ divisible by 3? The sum of the digits is,

$$3+9+5+6+0 = 11+5+6+0 = 16+6+0 = 22+0 = 22$$

and 22 is not divisible by 3, the closest multiple of 3 to 22 is $7 \times 3 = 21$. So, no, $39{,}560$ is not divisible by 3.

Is $877{,}349{,}596{,}691{,}899{,}982{,}199{,}209{,}202$ divisible by 3? With a bit of patience, we see that the sum of the digits is 153. Is 153 divisible by 3? The sum of its digits is $1+5+3 = 9$, which is 3×3, so, yes, 153 is divisible by 3 and, therefore, so is the original number.

Rule of 4

If the last two digits of the number are divisible by 4, then the entire number is divisible by 4. For example, $3{,}095{,}316 \div 4$ has no remainder because the last two digits, 16, are divisible by 4: $16 \div 4 = 4$.

Rule of 5

The rule of 5 is straightforward. If the number ends in 0 or 5, then it's divisible by 5.

Rule of 6

The rule for 6 is straightforward: if it's divisible by both 2 and 3, then it's divisible by 6.

Rule of 7

Seven is a bit of an oddball. The "rule" is actually a process:

- Take the last digit of the number and double it, i.e., multiply by 2.
- Subtract that answer from the remaining digits, meaning all the digits except the last one.

If the final number is zero or a multiple of 7, then the original number is also divisible by 7. Note, if you're not sure about the final number, you can repeat the process until you get a number divisible by 7 or 0.

Let's see this rule in action. Is 147 divisible by 7? First, take the last digit, 7, and double it: $7 \times 2 = 14$. Now, subtract this from the rest of the number without the final 7, which is 14,

$$14 - 14 = 0$$

the result is 0, so, yes, 147 is divisible by 7. Now, is 455 divisible by 7? Applying the steps, we first double 5 to get 10, then we subtract 10 from 45 to get $45 - 10 = 35$. Is 35 divisible by 7? Yes, since $7 \times 5 = 35$. Therefore, 455 is also divisible by seven.

Let's try a final example. What about 8639? Applying the process repeatedly gives,

$$863 - 18 = 845$$
$$84 - 10 = 74$$
$$7 - 8 = -1$$

so, no, 8639 is not divisible by 7. However, 8638 is divisible by 7. Use the steps of the rule to convince yourself that it is.

Rule of 8

The rule of 8 is like the rule of 4, except we look at the last three digits. For example, $3,095,320 \div 8$ has a remainder of zero because the last three digits are divisible by 8: $320 \div 8 = 40$. Because this rule requires dividing a three-digit number by 8, it's of limited utility unless the number we're wondering about is fairly large.

Rule of 9

The rule of 9 is like the rule of 3: if the sum of the digits is divisible by 9, then the number is divisible by 9.

Rule of 10

The rule of 10 is unique because 10 is the base of our place notation. *If the number ends in 0, then it's divisible by 10.*

Additionally, if the number ends in two zeros, like 1200, then it's divisible by 100. And, if it ends in three zeros, then it's divisible by 1000. Do you see the pattern? If the number ends in X zeros, then it's divisible by 1 followed by X zeros. Also, not only is the number divisible, the answer is what you get by chopping off X zeros on the right. A few examples,

$$
\begin{aligned}
1200 \div 10 &= 120 \\
1200 \div 100 &= 12 \\
230{,}000{,}000 \div 10{,}000{,}000 &= 23
\end{aligned}
$$

6.4.1 Section Summary

This section taught us a collection of rules we can use to quickly determine if a particular single-digit division problem will have a remainder or not. The rules come in handy from time to time, so they're worth remembering.

6.5 What's My Sign?

Division problems involving negative numbers follow the same rules as multiplication problems involving negative numbers. First, divide ignoring the signs of the numbers, then,

- If *either* of the numbers is negative, the answer is *negative*.
- If *both* of the numbers are negative, the answer is *positive*.

A few examples will suffice,

$$-456 \div 6 = -76$$
$$456 \div {}^-6 = -76$$
$$-456 \div {}^-6 = 76$$

6.6 Chapter Summary

This chapter introduced us to division, from division as repeated subtraction through the long division algorithm with single-digit divisors. Specifically, we learned that repeatedly subtracting a number from a larger number, while counting how many times we can subtract, tells us the number of times the number "goes into" the larger number, i.e., that the product of the number and the number of times we subtracted equals the larger number, with a possible remainder to add on. In addition, the number line helped us understand, visually, what the repeated subtractions were doing.

Next, we explored the essence of division and how place notation helps us understand how to decompose a number. This exploration led us to an idea, one that might help us implement an algorithm to find quotients and remainders.

Armed with the theory behind what division is actually doing, we were then in a place to learn the mechanics of long division using a single-digit divisor. We learned the recipe, and then glanced at an alternate viewpoint, one that more explicitly demonstrated what the recipe is doing.

Then, we learned quick rules to help us decide whether or not a number is evenly divisible by a particular single-digit number. Some of the rules were simple, like divisible by 10 or 5, while others required more effort, like divisible by 7.

We concluded the chapter by considering division problems involving negative numbers. We saw that such problems follow the same simple rules as multiplication problems with negative numbers.

This chapter was theory and mechanics of straightforward long division problems. To master long vision for arbitrary divisors, we need to spend a little more time with the algorithm. Practice makes perfect, or so the saying goes. I say practice makes you better. No one is perfect. Therefore, Chapter 7, will have us explore the long division algorithm with larger divisors.

6.7 Terms and Concepts

We introduced the following terms and concepts in this chapter.

dividend   The number being divided.

divisible   A number is divisible by another if there is no remainder after the division.

division   Division is the act of finding the largest multiple of the divisor that equals or is within one divisor's distance of the dividend. Division is the act of breaking up a number into the sum of a certain number of divisor-sized groups plus a possible remainder.

divisor   The number doing the dividing.

evenly divided   A number is evenly divided by another if there is no remainder after the division.

even number   A number which, when divided by 2, has no remainder. Also, a number whose final digit is 0, 2, 4, 6, or 8.

indeterminate   When applied to division, 0/0 is indeterminate because $0 \times a = 0$ is true for all numbers, a.

long division   The algorithm by which we can find the quotient and remainder for a division problem.

modulo   A first number modulo a second number returns the remainder, not the quotient. In other words, divide the first number by the second and keep only the remainder; discard the quotient.

multiple   A multiple of a number is that number times any other number. Division seeks to find a multiple, the product of the divisor and quotient, that is closest to the dividend.

odd number   Any number that is not even. When divided by 2, an odd number will have a remainder of 1. Also, any number whose final digit is 1, 3, 5, 7, or 9.

quotient   The answer to a division problem. The product of the quotient and the divisor, plus any remainder, equals the dividend.

remainder The part left over when subtracting the product of the quotient and the divisor from the dividend. The remainder is always less than the divisor. If the remainder is zero, then the dividend is divisible by the divisor.

undefined When applied to division, the answer to some number, b, other than zero, divided by zero, is undefined because the equation $0 \times a = b$ cannot be made true for any number, a.

6.8 Exercises

Exercise 1

Shortcuts are often most appreciated after experiencing the pain of doing things the long way. Therefore, solve the following division problems by using repeated subtraction. You'll notice that the single-digit long division algorithm won't help here. Is that crazed, maniacal laughter I hear?

$$73 \div 31 \qquad 168 \div 46 \qquad 151 \div 37$$
$$67 \div 42 \qquad 263 \div 48 \qquad 210 \div 26$$
$$193 \div 25 \qquad 171 \div 41 \qquad 268 \div 43$$

Exercise 2

Use the long division algorithm to solve the following single-digit division problems.

$$38034 \div 4 \qquad 20397 \div 6 \qquad 28182 \div 4 \qquad 7221 \div 2$$
$$51030 \div 7 \qquad 14275 \div 3 \qquad 86778 \div 2 \qquad 92351 \div 4$$
$$23252 \div 9 \qquad 34219 \div 3 \qquad 66420 \div 4 \qquad 13860 \div 4$$
$$56253 \div 9 \qquad 3971 \div 4 \qquad 57424 \div 2 \qquad 69334 \div 7$$

Think About It

The *Think About It* section of Chapter 5 introduced ancient Egyptian multiplication. The ancient scribes used a similar approach to divide. For example, to find $148 \div 4$, the scribes made a table with two columns. The first column started at 1 and each row multiplied the previous by 2. The second column started with the divisor, which was also multipled by 2 for each row. Therefore, to solve $148 \div 4$, the scribe made a table like so,

6.8. EXERCISES

1	4
2	8
4	16
8	32
16	64
32	128

Next, the scribe looked at the second column, working from bottom to top, adding each value of the second column to a running total if doing so gave a result less than or equal to the dividend. For example, here's the output where the rightmost column describes whether or not the row is included in the running total and why,

1	4	←	keep, $144 + 4 = 148$
~~2~~	~~8~~		ignore, $144 + 8 > 148$
4	16	←	keep, $128 + 16 = 144$
~~8~~	~~32~~		ignore, $128 + 32 > 148$
~~16~~	~~64~~		ignore, $128 + 64 > 148$
32	128	←	initial sum, 128

The scribe begins with 128, which is less than 148, so the scribe keeps it. The next row up is 64, but adding 64 to the total, currently 128, exceeds 148, so the scribe crosses out 64 and moves up to 32. Adding 32 to 128 is also too big, so it's crossed out, and the next row up is considered. Adding 16 to 128 gives 144 less than 148, so the 16 is kept, and the new running total is 144. Likewise, adding 8 is too big, but adding 4 is 148, so the 8 is ignored, and the 4 is kept. To find the quotient, the scribe adds the values of the first column for each included row, $1 + 4 + 32 = 37$. The quotient is correct since $37 \times 4 = 148$.

Dividing 148 by 4 leaves no remainder. What happens to the scribe's process when there is a remainder? How might the scribe know that the remainder isn't zero, and how might the scribe find it? As a practical example, repeat the process for $151 \div 4$, which has a remainder of 3. Where does the remainder show up, if at all?

Chapter 7

More Division

In general, we don't want more division in our lives, except when that division includes multi-digit divisors. In this short chapter, we apply the long division algorithm of Chapter 6 to problems where the divisor is no longer a single-digit number.

We'll ease into things by exploring two-digit divisors. Then we'll dive into the deep end and learn how to deal with divisors of arbitrary size. As we'll see, we only need to change the single-digit long division algorithm by a tiny bit in the end.

7.1 Two-Digit Divisors

Chapter 6 taught us that the goal of division is to find the largest multiple of the divisor that equals or falls just short of the dividend, the number being divided. The goal remains the same when the divisor has more than one digit; only the algorithm changes slightly. Let's see how by working through some examples with two-digit divisors. I think you'll see what is happening. Let's start with $59,397 \div 32$, which we set up exactly as we did in Chapter 6,

$$32 \overline{)59397}$$

How should we begin? For a single-digit divisor, we would look first at the 5, and, if that was less than the divisor, combine it with the 9 to work with 59.

We do much the same for the two-digit divisor, but we look at two digits of the dividend, at least initially. So, we look at 59 and ask: "what multiple of 32 is close to 59 without going over?" The answer is $1 \times 32 = 32$ because $2 \times 32 = 64$, and that's too large. Therefore, the first digit of the quotient is 1, and we write it over the 9 like so,

$$
\begin{array}{r}
1 \\
32\,\overline{\smash{)}59397} \\
\underline{32} \\
27
\end{array}
$$

Notice, I subtracted $1 \times 32 = 32$ from 59 to get 27. If this subtraction gives you a number equal to or greater than 32, you chose too small a multiplier. In that case, increase the quotient digit by one, multiply it by the divisor, and subtract again.

The next step of the long division algorithm says to bring down the next digit of the dividend, so, we bring the 3 down and ask: "what's the largest multiple of 32 that doesn't exceed 273?"

Here's where the bane of the long division algorithm with multi-digit divisors bites us. The multiplier will always be a single-digit number, but which one requires some guessing and some trial multiplications.

When I worked this problem myself, my first guess was 7, but that gives $7 \times 32 = 224$ and $273 - 224 = 49$, which is larger than 32, so 7 is too low. My next guess was 8 to get $8 \times 32 = 256$, which works. This leads to subtracting like so,

$$
\begin{array}{r}
18 \\
32\,\overline{\smash{)}59397} \\
\underline{32} \\
273 \\
\underline{256} \\
179
\end{array}
$$

with $273 - 256 = 17$ and the next digit, 9, already brought down.

Working with 179 leads to a trial multiplication of $5 \times 32 = 160$, which is as close as we can get, so subtracting and bringing down the 7 gives us,

$$
\begin{array}{r}
185 \\
32\,\overline{\smash{)}59397} \\
\underline{32} \\
273 \\
\underline{256} \\
179 \\
\underline{160} \\
197
\end{array}
$$

7.1. TWO-DIGIT DIVISORS

The final digit of the quotient is the multiple of 32 closest to 197. In this case, we can save ourselves a bit of work by noticing that $160 + 32 = 192$, and, since we just learned that $5 \times 32 = 160$, we know that $6 \times 32 = 192$, i.e., 32 more than 160. So, the final digit of the quotient is a 6,

```
         1856
   32)59397
      32
      ---
       273
       256
       ---
        179
        160
        ---
         197
         192
         ---
           5
```

Therefore, $59,397 \div 32 = 1856\mathrm{r}5$.

You may be wondering a bit about this process. For example, when working on a trial multiplication, as we did several times above, where do you write it, in a practical sense? The answer is on scrap paper, or near the actual problem as you work it. Do not, I implore you, attempt the trial multiplications in your head. Instead, write them somewhere, vertically, as we did in Chapter 5, and work them out step by step. Not writing things down is an invitation to frustration. Few of us can do mental math quickly and accurately, certainly not without considerable practice, to say nothing of memorizing a bag of tricks.

Moving on, then, lets solve $13,806 \div 91$. First, the setup,

```
   91)13806
```

Second, we ask how many times 91 goes into 13. The answer is zero, so we move right one digit of the dividend to consider 138. How many 91's in 138? The answer is one, so the first digit of the quotient is 1, and we place it above the 8, then subtract 91 and bring down the 0,

```
         1
   91)13806
      91
      ---
       470
```

Next, we ponder what times 91 is as close to 470 as possible. Let's guess 5. We calculate 5 × 91 = 455 on some scratch paper and see that it's within 91 of 470, so we have our next quotient digit,

$$\begin{array}{r} 15 \\ 91\overline{\smash{)}13806} \\ \underline{91} \\ 470 \\ \underline{455} \\ 156 \end{array}$$

Finally, to finish we realize that 1 × 91 = 91 is as close as we can get to 156, so a final multiplication by 1 and a subtraction gives,

$$\begin{array}{r} 151 \\ 91\overline{\smash{)}13806} \\ \underline{91} \\ 470 \\ \underline{455} \\ 156 \\ \underline{91} \\ 65 \end{array}$$

We end with a quotient of 151 and a remainder of 65. We haven't seen such a large remainder before, but that's okay. The only rule is that remainders are less than the divisor and 65 < 91, so we're fine, 13,806 ÷ 91 = 151r65.

Let's work an example that might be a bit confusing at first, but involves a situation that will happen from time to time. Let's find 84,126 ÷ 42,

$$42\overline{\smash{)}84126}$$

We begin as before, we ask how many times 42 goes into 84. The answer is 2 since 2 × 42 = 84. So, we subtract 2 × 42 to get,

$$\begin{array}{r} 2 \\ 42\overline{\smash{)}84126} \\ \underline{84} \\ 01 \end{array}$$

The subtraction produces a zero, and I've brought down the 1. So far, so good. Now, we ask how many times 42 goes into 1. The answer

7.1. TWO-DIGIT DIVISORS

is zero, meaning the next digit of the quotient is 0. At this point, we could continue the algorithm multiplying 42 by 0 to get 0, then subtract 0 leaving 1, and bring down the next digit, the 2, to get 12. Or, we could put 0 as the next digit of the quotient and immediately bring down the 2. We saw a situation like this in Chapter 6. At this point, the problem looks like this,

$$
\begin{array}{r}
20 \\
42\overline{\smash{)}84126} \\
\underline{84} \\
012
\end{array}
$$

To continue, we see that 12 is still less than the divisor of 42. So, we again multiply by 0, i.e., make the next quotient digit 0, and bring down the 6,

$$
\begin{array}{r}
200 \\
42\overline{\smash{)}84126} \\
\underline{84} \\
0126
\end{array}
$$

We're finally in a place to move forward. How many times does 42 go into 126? Three, $3 \times 42 = 126$. So, we finish by subtracting 126,

$$
\begin{array}{r}
2003 \\
42\overline{\smash{)}84126} \\
\underline{84} \\
0126 \\
\underline{126} \\
0
\end{array}
$$

There is no remainder, so $84,126 \div 42 = 2003$.

If the value found after a subtraction and the bringing down of the next dividend digit is less than the divisor, as was the case here, the corresponding quotient digit is 0, and we bring down successive digits of the dividend until we have a value that is larger than the divisor, or we run out of digits in the dividend.

Let's try one more example, an extreme one. Let's find $84,001 \div 42$. First, the setup,

$$
42\overline{\smash{)}84001}
$$

followed by the first step answering the question, "how many 42's in 84?"

$$42\overline{)84001}\atop{\underline{84}}\atop00$$ with quotient 2

As before, the result of the subtraction is zero, and I brought the next 0 down from the dividend. The resulting value is still 0, so, as before, we bring down the next 0 from the dividend. However, this is still 0, so we continue and bring down the final digit of the dividend, the 1,

$$42\overline{)84001}$$ with quotient 200, subtract 84, remainder 0001

To complete the problem, we are compelled to use 0 as the final digit of the quotient because there are no more digits of the dividend to bring down. Therefore, we write,

$$42\overline{)84001}$$ with quotient 2000, remainder 1

The end result is $84,001 \div 42 = 2000\text{r}1$.

The main difference between long division with single-digit divisors versus two-digit divisors was the first step, the part where we find the first set of dividend digits that is equal to or larger than the divisor. After that, we follow the recipe bringing down one digit of the dividend at a time until we are done. Let's finish our exploration of division, at least with integers, by considering arbitrary divisors.

7.1.1 Section Summary

This section introduced long division with two-digit divisors. We learned that the first step needs to be modified to consider as many digits of the dividend as necessary to have a value equal to or larger than the divisor. For a two-digit divisor, this means the first two digits, or three at most, which we saw with our first worked example. Why three at most? Because the largest two-digit divisor is 99, and any three-digit number must be larger than that. We worked through several examples and encountered situations where we need to continue to bring down digits of the dividend until we have a value larger than the divisor.

The progression from single-digit to two-digit divisors has, hopefully, helped you gain a deeper insight as to what the division algorithm is doing. As always, the book's website has many worked example problems. Please avail yourself of them until you are comfortable with the process and different situations that might arise.

7.2 Arbitrary Divisors

Let's wrap up our exploration of long division by considering a few problems with arbitrary divisors. We'll revisit long division in Chapter 9 when we investigate decimal numbers, but for now, we know all we need to know.

Our first example uses a three-digit divisor. Based on our observation with two-digit divisors, we expect to examine at most four digits of the dividend to start the process. I.e., if there are n digits in the divisor, then we need at most $n + 1$ digits of the dividend to get started. Therefore, let's find $378,014 \div 251$.

I'll present the steps in order. Your task is to read through the steps below and pull out what happened to move from one to the next. I'll make some comments afterward as to why the steps are what they are.

$$251\overline{)378014}$$

$$\begin{array}{r} 1 \\ 251\overline{)378014} \\ \underline{251} \\ 1270 \end{array}$$

```
            150
    251)378014
        251
        1270
        1255
         1514
```

```
           1506
    251)378014
        251
        1270
        1255
         1514
         1506
            8
```

To start the problem, we look at the first three digits of the dividend, 378, which is greater than the divisor of 251. Therefore, the first digit of the quotient goes above the 8 of the dividend.

Subtracting 1×251 from 378 leaves 127, which becomes 1270 when the next digit of the dividend, the 0, is brought down. I guessed that 5×251 would be close to 1270. Working it out on scrap paper validated my guess, $5 \times 251 = 1255$. Therefore the second digit of the quotient is 5. Subtracting leaves 15.

Bringing down the 1 gives us 151, which is less than 251, so the 4 comes down as well. We're left with finding the digit that multiplied by 251 is as close to 1514 as possible. The answer is 6 since $6 \times 251 = 1506$. Therefore, the full quotient is 1506, with a remainder of 8.

The amount of effort required to complete a long division problem depends, primarily, on the pain experienced performing the trial multiplications. For example, here are the steps to find $12{,}345{,}678 \div 9999$,

```
    9999)12345678
```

```
              1
    9999)12345678
         9999
         23466
```

7.2. ARBITRARY DIVISORS

$$
\begin{array}{r}
12 \\
9999{\overline{\smash{\big)}\,12345678}} \\
\underline{9999} \\
23466 \\
\underline{19998} \\
34687
\end{array}
$$

$$
\begin{array}{r}
123 \\
9999{\overline{\smash{\big)}\,12345678}} \\
\underline{9999} \\
23466 \\
\underline{19998} \\
34687 \\
\underline{29997} \\
46908
\end{array}
$$

$$
\begin{array}{r}
1234 \\
9999{\overline{\smash{\big)}\,12345678}} \\
\underline{9999} \\
23466 \\
\underline{19998} \\
34687 \\
\underline{29997} \\
46908 \\
\underline{39996} \\
6912
\end{array}
$$

In this case, the pain is from calculating the trial multiplications of 9999. Other than that, the problem is straightforward.

Let's end this section with one more example, one that turns out to be quite simple to solve. Let's see $11011011 \div 1001$,

$$
\begin{array}{r}
11000 \\
1001{\overline{\smash{\big)}\,11011011}} \\
\underline{1001} \\
1001 \\
\underline{1001} \\
0011 \\
\underline{0} \\
11
\end{array}
$$

This example is straightforward because all trial multiplications are either zero or one times the divisor. Therefore, $11011011 \div 1001 = 11000\text{r}11$.

What happens when the divisor is larger than the dividend? For example, what is 818 ÷ 1024? Let's set up the problem,

$$1024\overline{)818}$$

To solve it, we need to examine the first four digits of the dividend because the divisor has four digits. However, the dividend is only three digits long; it's less than the divisor, so the quotient is zero,

$$\begin{array}{r} 0 \\ 1024\overline{)818} \\ \underline{0} \\ 818 \end{array}$$

We can write 818 ÷ 1024 = 0r818, meaning all of the dividend is considered the remainder. Does this mean we cannot divide a smaller number by a larger? Actually, no, we can, and we'll encounter approaches to doing so in Chapter 8 on fractions and again in Chapter 9 when we explore decimal numbers. To whet your appetite, the answer to 818 ÷ 1024 is a number that is less than one but greater than zero.

7.2.1 Section Summary

In this section, we explored division with arbitrary-sized divisors. The general rule to observe is that we begin with enough of the dividend to be at least as large as the divisor. This means that if the divisor has n digits, we need at most $n + 1$ digits of the dividend to start long division.

We also learned that the trial multiplications while working out each quotient digit can be rather painful as the divisor gets larger and larger. That's just the way it goes. Fortunately, we rarely encounter situations where we need to divide a large integer by another large integer, so we need not suffer too often.

7.3 Chapter Summary

This chapter did not introduce anything new beyond the slight adjustment to the long division algorithm necessary to get started when the divisor has more than one digit. We explored the consequences of this adjustment, first with two-digit divisors, and then with divisors of arbitrary size. In all cases, the body of the long division algorithm remained the same, as did our overall goal: to find the largest multiple of the divisor that is equal to or as close as possible to the dividend. Any remaining difference is, well, the remainder, as expected.

Our exploration of the big four arithmetic operations is now complete, at least for integers. Our next stop is the world of fractions, the topic of Chapter 8. So strap in and prepare for liftoff.

7.4 Exercises

Exercise 1

Use the long division algorithm to solve the following division problems.

$26776 \div 436$	$96049 \div 104$	$24874 \div 87$	$23662 \div 276$
$77328 \div 68$	$71389 \div 992$	$97731 \div 249$	$86260 \div 555$
$67737 \div 43$	$70930 \div 279$	$29685 \div 439$	$67337 \div 751$
$12510 \div 717$	$92300 \div 388$	$38548 \div 956$	$34884 \div 466$

Think About It

All the numbers that evenly divide another number are known as the factors of the number. For example, the factors of 12 are 1, 2, 3, 4, 6, and 12 because,

$$1 \times 12 = 12$$
$$2 \times 6 = 12$$
$$3 \times 4 = 12$$
$$4 \times 3 = 12$$
$$6 \times 2 = 12$$

Likewise, the factors of 10 are 1, 2, 5, and 10 because $1 \times 10 = 10$ and $2 \times 5 = 10$. However, the only factors of 7 are 1 and 7 since $1 \times 7 = 7$ is the only way to write 7 as the product of two integers. Numbers that are only divisible by 1 and themselves have a special name, they are called *prime numbers*, and mathematicians have been studying them and their properties for centuries. Numbers that are not prime are known as *composite numbers*. All composite numbers can be written as the product of prime numbers; this is known as the *prime factorization* of a number.

There are an infinite number of primes. As of this writing, the largest known prime number has nearly 25 million digits. Primes are crucial to secure digital information transfer, which is often based on the product of two large prime numbers.

One method for finding prime numbers dates from antiquity. We know it today as the *Sieve of Eratosthenes*. Eratosthenes was an ancient Greek mathematician, though the technique was only attributed to him centuries after his death.

For example, to find all the prime numbers less than 20, first write the numbers in sequence starting with 2,

$$\begin{array}{cccccccccc} & 2 & 3 & 4 & 5 & 6 & 7 & 8 & 9 & 10 \\ 11 & 12 & 13 & 14 & 15 & 16 & 17 & 18 & 19 & 20 \end{array}$$

Then, beginning with 2, start counting by twos crossing off every number you land on. This means cross off all multiples of 2. Doing so gives,

$$\begin{array}{cccccccccc} & 2 & 3 & \cancel{4} & 5 & \cancel{6} & 7 & \cancel{8} & 9 & \cancel{10} \\ 11 & \cancel{12} & 13 & \cancel{14} & 15 & \cancel{16} & 17 & \cancel{18} & 19 & \cancel{20} \end{array}$$

Then, proceed to the next number not already crossed off, which is 3, and repeat, but now count by threes,

$$\begin{array}{cccccccccc} & 2 & 3 & \cancel{4} & 5 & \cancel{6} & 7 & \cancel{8} & \cancel{9} & \cancel{10} \\ 11 & \cancel{12} & 13 & \cancel{14} & \cancel{15} & \cancel{16} & 17 & \cancel{18} & 19 & \cancel{20} \end{array}$$

If a number is already crossed off, that's okay; count from it anyway.

Move to the next available number, here 5, then count by fives. Continue, each time counting by whatever number is still available, until all numbers have been tested. For the example, we end with,

$$\begin{array}{cccccccccc} & 2 & 3 & \cancel{4} & 5 & \cancel{6} & 7 & \cancel{8} & \cancel{9} & \cancel{10} \\ 11 & \cancel{12} & 13 & \cancel{14} & \cancel{15} & \cancel{16} & 17 & \cancel{18} & 19 & \cancel{20} \end{array}$$

because counting by 5 would mark 15 and 20, already marked, whereas counting by 7 would mark 14, and counting by 11 is already past the limit of 20. All the numbers remaining are the prime numbers less than 20: 2, 3, 5, 7, 11, 13, 17, and 19.

What are all the primes less than or equal to 100?

Chapter 8

Fractions

Jonna Mendez is a real-life master of disguise. She and her late husband, Tony, spent decades working for the US Central Intelligence Agency inventing means by which spies could disguise themselves, often within seconds. While not as exciting, *fractions*, the topic of this chapter, are also masters of disguise; they exist as a concept but have many faces. Our task, should we choose to accept it, is to unmask fractions learning to manipulate them via addition, subtraction, multiplication, and division. Along the way, we'll encounter mixed numbers, another shadowy type related to fractions.

To be specific, we'll learn that fractions are parts of a whole. We'll then learn where fractions live on the number line. Next, we'll see through fractions' many disguises to reduce them to their most basic form.

Representing and reducing fractions prepares us for multiplication and division. For fractions, multiplication and division are simpler than addition and subtraction. Go figure. After multiplication and division come addition and subtraction. One type of fraction is easy to add and subtract. Most general fractions require more effort. A quick look at fractions as ratios follows. Negative fractions are a thing, so we consider them next.

We end the chapter by contemplating scandalous mixed numbers and improper fractions. We'll learn what they are, how to use them, and how to disguise one as the other and vice versa.

We have a mole in our organization. Let's get to work uncovering it.

This is a lengthy chapter packed with new concepts. You may have some difficulty following them all on a first reading. If so, that's perfectly fine. I expect multiple passes through the chapter will be necessary. I recommend a quick reading of the chapter, then try some of the exercises. If the exercises are not making sense, go back and review the chapter again, paying close attention to each step of the worked examples. The minimum goal of the chapter is for you to be comfortable doing the exercises.

The spy metaphor for fractions isn't just for fun. Fractions are a general concept, and that means they can be thought of in multiple ways. The solution for confusion is dogged struggle with the text and review of the exercises and their answers. To help you, the book's website, as always, contains many practice problems with solutions.

8.1 Fractions Are...

Fractions are many things. Let's cover some of what they are here. By the end of this section, we'll know what fractions are in terms of parts of a whole, as labels on the number line, and as a new kind of number.

8.1.1 Parts of A Whole

If I have a chocolate bar and break it into four pieces, I'll most likely wave them in your face as I eat them, one by one. After all, it's chocolate. Nice people, however, would probably give you a piece of the candy bar while keeping some for themselves. Let's make a picture of me being nice and sharing my chocolate bar with you. It's only an illustration because I'd never actually do something so foolish where chocolate is concerned. Here's the picture,

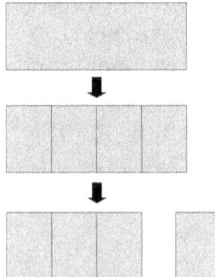

8.1. FRACTIONS ARE... 161

Van Gogh, I'm not, I admit it, but work with me on this. The *whole* chocolate bar is on the top. Next, we have the chocolate bar cut into four *equal-sized* pieces. The fact the pieces are all the same size matters. Finally, the chocolate bar is now split with three pieces on the left for me and one piece on the right for you.

We've just turned a whole into parts. We turned the chocolate bar into fractions because *fractions are parts of a whole*. In this case, we turned the whole into four parts. Splitting a whole into four equal-sized parts is so common that the parts have a special name: *quarters*. That's why there are four quarters to a dollar.

Okay, how many quarters do you have, and how many do I have? You have one-quarter of the chocolate bar and I have three-quarters of the bar. Combined, they are the entire bar, four quarters. Let's write "three quarters" and "one quarter" as fractions,

$$\frac{3}{4} \quad \frac{1}{4}$$

Another way to write these fractions is with the slash: 3/4 and 1/4. Wait a second. In Chapter 6, we learned that slash was a way to indicate division. That's true; fractions are division problems in disguise. However, these are division problems we won't work out (whatever that means, exactly) because, as we'll see below, we can work with the problems as they are.

The parts of a fraction have names. It's best to learn them now; we'll use them extensively throughout the chapter,

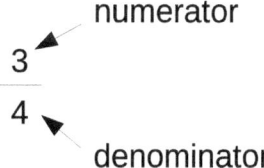

The *denominator* tells us how many pieces the whole has been split into. In this case, four pieces. The *numerator* tells us how many of those pieces we have, here three. The fraction 1/4 means the whole has been split into four pieces, and we have one of them. Likewise, the fraction 11/13 means the whole is in 13 pieces, and we have 11 of them, and so on. Therefore, a fraction is a part of a whole.

Before we go any further, we need to spend a bit of time talking about how to say fractions, meaning how to speak about them out loud. In Chapter 1, we learned the proper way to say numbers out loud. There are rules for fractions as well. For example, here are the fractions for splitting something into four parts and how we should say them out loud,

$\frac{1}{4}$ "one fourth"

$\frac{2}{4}$ "two fourths"

$\frac{3}{4}$ "three fourths"

$\frac{4}{4}$ "four fourths" (or just "one")

Notice, when the numerator and denominator are equal, we have one, the whole thing. When we have all the pieces, we have the thing, and there's only one thing, so any fraction where the numerator and denominator are equal is the number 1. It's essential to understand and remember this; we'll see it again many times throughout the chapter.

As for naming fractions, in most cases, we say the denominator with "ths" appended. For example, 11/13 is "eleven thirteenths." However, human language isn't entirely rational, so when the denominator is 2, we use "half," therefore, 1/2 is "one half." Likewise, for a denominator of three, we use "thirds" so 1/3 is "one third" and 2/3 is "two thirds." This business extends to larger denominators ending in three as well. So, 21/23 is "twenty-one twenty-thirds" and so forth.

8.1.2 On The Number Line

Where are fractions on the number line? The answer is between the integers. Let's see why. Consider this number line,

I marked 0 and 1. I also put marks between 0 and 1 so that the range is divided into eight pieces. Here are tally marks again; we never really get away from them. What do the marks represent? If a fraction is a part of a whole split into equal-sized pieces, and there

8.1. FRACTIONS ARE...

are eight of them here between 0 and 1, then these marks represent the fractions with a denominator of 8. All we need to do to label the marks is write the fractions in order like this,

Notice, the fraction 1/8 has covered one of the eight pieces going from 0 to 1. Imagine an arrow beginning at 0 and ending at 1/8. Similarly, the arrow from 0 to 7/8 covers seven of the eight pieces. The numerator "numbers" the denominator, which is why it's called a numerator. If you're wondering whether the numerator can be bigger than the denominator, it can. We'll get to fractions like this later in the chapter as well (so many promises!)

What else can we glean from this number line? Above, I said that any fraction where the numerator and denominator are the same equals 1; therefore, we know that 8/8 = 1. This means that all fractions like 3/8 are less than 1 but greater than 0. Fractions exist on the number line between 0 and 1. This is true so long as the numerator is less than the denominator.

Here's another number line. What labels go on each of the marks?

The range is split into four pieces; therefore, the denominator is 4, i.e., these are quarters. To label the marks, then, we write each numerator in turn,

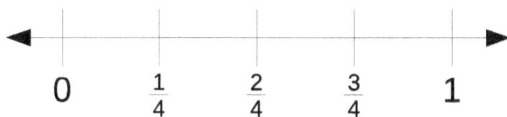

Finally, here's a number line where the range is split into two pieces, i.e., in half,

which we mark as,

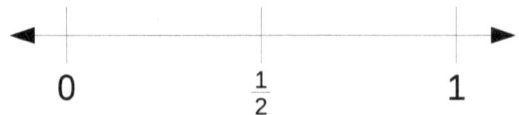

The label 1/2 makes sense. The first part, beginning at 0, covers half the distance to 1, so it is 1 part out of the 2.

You may have noticed a few things about these number lines. One is that the marks look much like what you see on a ruler. There's a reason for that, one that we'll explore more later in the chapter. But, yes, a ruler is a number line where the marks represent fractions from one number to the next. Another thing you may have noticed is that some of the marks on the number lines divided into 8, 4, and 2 pieces line up. We'll return to that shortly as well. For now, the takeaway is that fractions live on the number line between the integers, and fractions where the numerator is less than the denominator live between 0 and 1.

The next section introduces fractions as a new kind of number. It's more theoretical, so you may wish to skip it at first, then come back to it after completing the chapter.

8.1.3 Rational (Optional)

Fractions have another name. To mathematicians, fractions, meaning numbers that can be written as an integer divided by another integer that is not zero, are known as *rational* numbers. The word "rational" comes from the same root as "ratio" because a fraction is a ratio between two integers. We'll work with ratios later in the chapter, so no worries if the word is unfamiliar.

Rational numbers are, as a group, represented by $\mathbb{Q}$, for "quotient," again alluding to the fact that fractions are division problems in disguise. The rationals are a new kind of number. What is their relationship to the other types of numbers we already know, the integers, whole numbers, and counting numbers?

In Chapter 1 we saw a diagram showing that all counting numbers are inside the group (set) of whole numbers and that all whole numbers are inside the set of integers. Consider the formal mathematical definition of a rational number: a rational number is an integer, p, divided by another integer, q where q cannot equal zero (why?).

8.1. FRACTIONS ARE...

What happens if we fix $q = 1$? We have all the rational numbers where the denominator is 1, but, if $p/1$ is $p \div 1$, what is the quotient when dividing a number by 1? The answer is the number itself, so $p/1 = p$ for any integer p. I.e., the integers are a special group of rational numbers and the set of rational numbers holds all the integers plus numbers that are rational but not integers. Therefore, we need to add another outer doll to our collection of Russian dolls,

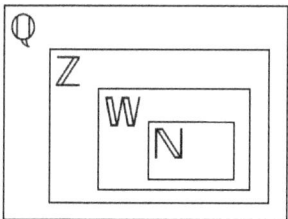

The diagram tells us that the rational numbers contain all integers, the integers contain all whole numbers, and the whole numbers contain all counting numbers (also called natural numbers). Recall, counting numbers are positive integers beginning with 1 and whole numbers are the counting numbers and 0.

In truth, while rational numbers exist, as much as any number exists, writing 3/8 to label a specific rational number, i.e., the one that is as far from 0 as the sum of three segments when the range from 0 to 1 is divided into eight pieces, is a necessary evil. How else might we label this number? That's a rhetorical question; if there were a nicer way to label the rationals, we'd be using it. My complaint is due to the fact that rational numbers have many names for the same number because many division problems end up at the same place on the number line. You may have already noticed this from the number lines above, but 1/2, 2/4, and 4/8 are all labels for the same place between 0 and 1 on the number line. Every rational number has an infinite number of labels, disguise after disguise. Who knew that math could be so sneaky?

8.1.4 Section Summary

This section told us that fractions are parts of a whole. This is the most fundamental way to think about fractions. We then learned where fractions live on the number line, in the space between the integers. Finally, we learned that fractions are a new kind of number, the rational numbers, and that all the numbers we know of already are really special types of rational numbers.

8.2 Fractions Undercover

Fractions have many faces. In this section, we learn why this is so followed by how to reduce a fraction to uncover its true face.

8.2.1 Fractions Have Aliases

The section above introduced us to three number lines. Here they are again, one on top of the other,

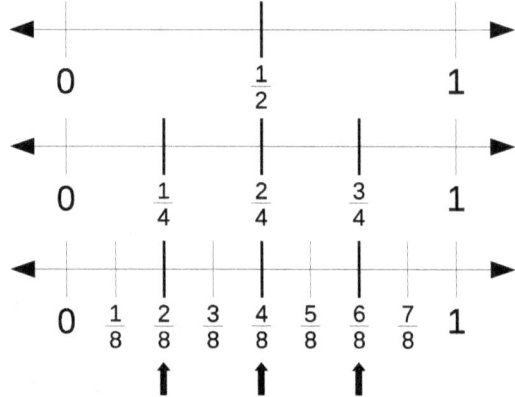

Do you notice anything in common between the three? The big arrows are a clue, of course. The same place on the number line, the same number, is called by different names for the locations marked. For example, the middle is called 1/2, then 2/4, and finally 4/8. Likewise, the left and right of the middle are labeled 1/4 and 2/8 or 3/4 and 6/8. Fractions have multiple names, multiple ways to write the same number. Each fraction has, in fact, an infinite number of names.

Let's write the middle fractions like this,

$$\frac{1}{2} = \frac{2}{4} = \frac{4}{8}$$

and let's look across at the numerators and denominators. The sequence of numerators is 1, 2, 4 and the sequence of denominators is 2, 4, 8. To go from one numerator to the next, we multiply by 2. And, to go from one denominator to the next, we also multiply by 2. This pattern continues, so we can multiply the numerator and denominator of 4/8 by 2 to get 8/16 as yet another name for the same number. There is nothing special about multiplying by 2. We can

8.2. FRACTIONS UNDERCOVER 167

multiply the numerator and denominator by 17 if we wish to get 68/136 as still another name. What matters is the ratio between the numerator and the denominator. We'll talk about ratios in time.

I'm going to let you in on a little secret. *To multiply a fraction by another, multiply the numerators and then multiply the denominators.* Let's see how this might explain the many names for the same number. I'm going to write,

$$\frac{1}{2} \times \frac{2}{2} = \frac{2}{4}$$

I followed the rule for multiplying fractions, the numerator is $1 \times 2 = 2$ and the denominator is $2 \times 2 = 4$. However, look at the fraction I multiplied 1/2 by. It's 2/2. We know that $2/2 = 1$, so what I really did was multiply 1/2 by 1. Therefore, $1/2 = 2/4$. The alternate names for fractions come about when we multiply by 1 wearing different disguises, by a different alias for 1.

Let's try another,

$$\frac{2}{3} \times \frac{5}{5} = \frac{10}{15}$$

Here, I used the 5/5 alias for 1. This means that $2/3 = 10/15$ because, again, multiplying a number by 1 gives you the number back.

There are two things to notice. First, multiplying a fraction by a specific alias for 1 gives you a new fraction that is really the first fraction in disguise. Second, even though there are an infinite number of names for a fraction, there is one base name, one name that is not some alias of 1 times another fraction. This is the name that we want to use because it is unique. For the number lines above, the middle line's base name is 1/2, there is no fraction times any alternate name for 1 that gives us 1/2 as the answer. The following section shows us how to take a fraction and find this base name.

8.2.2 Reducing Fractions

The proper way to talk about a fraction's base name is to say that the fraction is in *lowest terms*. The process of finding the lowest terms for a fraction is known as *reducing* the fraction. Let's see how.

I said above that 10/15 is another name for the fraction 2/3, two-thirds. To reduce a fraction, we need to find which alias of 1 the lowest term fraction was multiplied by to give us the fraction we're contemplating. This is equivalent to finding the largest number that divides both the numerator and denominator. In other words, we need to find the *greatest common divisor* or GCD of the numerator and denominator. Once we have the GCD, we have what we need to reduce the fraction.

To find the GCD, we need to know the factors of the numerator and denominator. The factors of a number are the collection of numbers in the products that give that number. For example, the factors of 10 are 1, 2, 5, and 10 because $1 \times 10 = 10$ and $2 \times 5 = 10$ and no other pair of integers when multiplied gives 10. Likewise, for 15, the factors are 1, 3, 5, and 15. Let's write the factors of 10 and 15 in order,

factors of 10: 1, 2, **5**, 10
factors of 15: 1, 3, **5**, 15

Notice, I highlighted 5, as that is the largest factor 10 and 15 have in common. Therefore, the GCD of 10 and 15 is 5. Okay, fine, now what? Now, we *divide the numerator and denominator by the GCD*. This gives us a new fraction, the one that is in lowest terms. In this case, $10 \div 5 = 2$ and $15 \div 5 = 3$ so the new fraction is 2/3, meaning 10/15 is really 2/3 in disguise.

Let's work another example; let's find 176/208 in lowest terms. To do so, we need to know the factors of 176 and 208. How do we get the factors of a number? Usually by trial and error. Here's where the single-digit division rules of Chapter 6 can help us. Both 176 and 208 are even, so 2 is a factor. The sum of the digits of 176 is 14 and of 208 is 10. Neither 14 nor 10 are divisible by 3 or 9, so 3 and 9 are not factors of either number, nor is 6 as a number is divisible by 6 only if it's divisible by both 2 and 3.

A nontrivial set of trial divisions tells us that the factors of 176 are 1, 2, 4, 8, 11, 16, 22, 44, 88, and 176. For 208 we get factors of 1, 2, 4, 8, 13, 16, 26, 52, 104, and 208. The greatest factor in common is 16, so the GCD is 16. Therefore, to get 176/208 in lowest terms, divide the numerator and denominator by 16 to get $176 \div 16 = 11$ and $208 \div 16 = 13$ meaning the lowest terms fraction is 11/13.

What happens if the factors of the numerator and denominator have no factor in common? In that case, the fraction is already in lowest terms, there is no alias of 1 that multiplied a simpler fraction.

In practice, working through the factors of numbers is a pain. In most cases, we can use a faster approach. Here's what I do. I start with 2 and ask if the numerator and denominator can be divided by 2. If so, I do that and keep successively dividing by 2 until I can't anymore. For example,

8.2. FRACTIONS UNDERCOVER

$$\frac{176}{208} = \frac{88}{104} = \frac{44}{52} = \frac{22}{26} = \frac{11}{13}$$

The original fraction is on the left; each fraction moving to the right is the previous fraction, where the numerator and denominator are divided by 2. On the right, we eventually get to a fraction that can no longer be divided by 2. In this case, 11 and 13 are prime, they have no factors, so we are done. In other cases, I would then ask if I can divide by 3; if not, then 5, 7, 11, etc.

The list of numbers I'm using here isn't arbitrary; I'm using the primes. The reason why using the primes will eventually get you to the lowest terms has to do with the fact that every number can be written as the product of a series of primes. Objectively, the benefit of this approach is that the numerator and denominator are smaller and smaller as we move to the right, so finding the factors gets easier and easier.

Let's try another example, let's reduce 144/324. Again, we start with successive divisions of the numerator and denominator by 2 until we can't anymore,

$$\frac{144}{324} = \frac{72}{162} = \frac{36}{81}$$

So far, so good. But, we are not done. How do I know? Because both 36 and 81 are divisible by 3, so we can continue now dividing by 3,

$$\frac{36}{81} = \frac{12}{27} = \frac{4}{9}$$

At this point, we are at lowest terms because 4 and 9 have no factors in common. Therefore, 144/324 = 4/9.

Reducing fractions is essential when doing arithmetic with them. A few multiplications and additions can quickly create correct fractions with large numerators and denominators if we don't reduce them. The usual process is to do the operation, then reduce the answer to lowest terms before proceeding.

To see what I mean, let's multiply 26/34 by 3/8. Both of these fractions are already in lowest terms. Multiplying them gives us,

$$\frac{26}{34} \times \frac{3}{8} = \frac{78}{264}$$

and, reducing 78/264 gives us,

$$\frac{78}{264} = \frac{39}{132}, \quad \frac{39}{132} = \frac{13}{44}$$

where first we divide numerator and denominator by 2, then by 3, to get the reduced fraction. The final answer is,

$$\frac{26}{34} \times \frac{3}{8} = \frac{13}{44}$$

The answer might be confusing at first, as we said to multiply two fractions, we multiply the numerators and then the denominators, but 26×3 isn't 13. The step implied above reduces the answer, so while it might look strange, the answer is correct.

Let's summarize the process to reduce a fraction to lowest terms:

1. Begin with 2 as the current divisor.

2. Divide the numerator and denominator by the current divisor repeatedly until one or the other is no longer divisible by the current divisor.

3. Move to the next prime as the current divisor. The first few primes are: 2, 3, 5, 7, 11, 13, 17, 19.

4. If the resulting fraction can no longer be divided by any prime, it's in lowest terms.

Let's apply the algorithm to one last example, 17640/22680. Begin with 2:

$$\frac{17640}{22680} = \frac{8820}{11340} = \frac{4410}{5670} = \frac{2205}{2835}$$

then move to 3,

$$\frac{2205}{2835} = \frac{735}{945} = \frac{245}{315}$$

and now 5,

$$\frac{245}{315} = \frac{49}{63}$$

and, finally, by 7,

$$\frac{49}{63} = \frac{7}{9}$$

to see that 17640/22680 = 7/9. Whew! Have no fear; the exercises will not ask you to deal with such large numerators and denominators, but if you had to, you could. The numerator and denominator of 17640/22680 are both large, five-digit numbers. Is the fraction itself large? No, we just calculated that it's merely another name for 7/9, which is less than 9/9 = 1.

8.2.3 Section Summary

The objectives of this section were twofold. First, it was to understand why fractions have multiple names. We learned an infinite number of fractions could label the same number because we can multiply a fraction by any of an endless number of aliases for 1. We then learned that despite this, each fraction has a unique lowest terms version, which is not multiplied by any alias for 1.

The second objective of this section was to learn how to reduce a fraction to find its lowest terms representation. We learned that dividing the numerator and denominator by the greatest common divisor, the largest factor of each in common, will immediately reduce the fraction to lowest terms. We also learned that finding this value, in practice, is painful. In its place, we learned a simple algorithm to keep dividing numerator and denominator by ever-larger primes until we can't anymore. At that point, the fraction is in lowest terms.

Multiplying fractions is straightforward. Dividing fractions is nearly as easy. Let's learn how now.

8.3 Dividing Fractions

We covered how to multiply fractions above. To divide them is equally straightforward, but it's also an opportunity to introduce the concept of a reciprocal, so we'll do that first, then use the idea to divide fractions.

8.3.1 Reciprocals

A fraction is one integer divided by another: p/q for two integers, p and q, whatever they are ($q \neq 0$). The *reciprocal* of a fraction is the number you get when you flip the numerator and denominator. So, if p/q is the fraction, then the reciprocal is q/p, as long as the number you pick for p isn't zero. For example, here some fractions and their reciprocals,

$$\frac{2}{3} \rightarrow \frac{3}{2}$$

$$\frac{11}{13} \rightarrow \frac{13}{11}$$

$$\frac{17}{8} \rightarrow \frac{8}{17}$$

$$\frac{123}{456} \rightarrow \frac{456}{123}$$

$$\frac{1}{6} \rightarrow \frac{6}{1} = 6$$

I'm confident you get the idea. Notice, the reciprocal of a fraction with a numerator of 1 (called a *unit fraction*) is an integer, the denominator alone.

So, why mention reciprocals now? We'll use them for dividing fractions, but I want you to notice two things before we do that. First, what happens if we multiply a fraction by its reciprocal? For example,

$$\frac{2}{3} \times \frac{3}{2} = \frac{6}{6} = 1$$

Multiplying a fraction by its reciprocal always gives 1 as the answer. In formal math language, the reciprocal of a fraction is its *multiplicative inverse*, the number that gives you 1 when you multiply the fraction by it.

Second, did you notice a new type of fraction above? The reciprocal of 2/3 is 3/2. This is the first time we've seen a fraction where the numerator is greater than the denominator. This is perfectly valid, but what does it mean?

In this case, 2/3 means the interval from 0 to 1 has been divided into 3 parts and we have two of them,

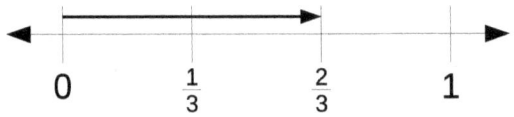

The fraction 3/2 means that we have divided 0 to 1 into 2 parts, and we have 3 of them. On the number line, this means we are coving a

8.3. DIVIDING FRACTIONS

distance from 0 that is three segments, where each segment is half the distance from 0 to 1,

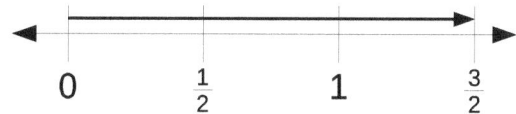

The fraction 3/2 is an *improper fraction*. Improper fractions are scandalous fractions that allow their numerators to be greater than their denominators. As the number line above shows, improper fractions are larger than 1. The fraction 3/2 covers the range from 0 to 1 and halfway again to 2.

What about 4/2 as an improper fraction? Here, we make use of the fact that fractions are division problems, so $4/2 = 4 \div 2 = 2$. I.e., adding four halves together gives us two wholes.

Let's use reciprocals to learn how to divide fractions.

8.3.2 Dividing By Multiplying

Dividing two fractions is easy: *first, find the reciprocal of the second fraction, then multiply the first fraction by it*. For example,

$$\frac{2}{3} \div \frac{3}{4} = \frac{2}{3} \times \frac{4}{3} = \frac{8}{9}$$

and,

$$\frac{5}{7} \div \frac{1}{2} = \frac{5}{7} \times \frac{2}{1} = \frac{10}{7}$$

In this case, the reciprocal of 1/2 is 2/1 = 2, but when multiplying a fraction, its often easier to leave the denominator of 1. This holds for multiplying a fraction by an integer directly,

$$\frac{5}{7} \times 2 = \frac{10}{7}$$

where we imagine a denominator of 1 for the 2, i.e., 2/1, or remember to multiply the numerator alone if the other number is an integer.

That's all there is to division of two fractions: multiply by the reciprocal of the second. Now, let's change gears a bit and consider fractions in a new way, as ratios.

8.3.3 Section Summary

This brief section first introduced the idea of a reciprocal as flipping the numerator and denominator, then demonstrated that multiplying a fraction by its reciprocal always equals 1. Next, we used the reciprocal to learn that division of fractions is multiplication by the reciprocal of the second. Along the way, we learned about improper fractions, those fractions where the numerator exceeds the denominator. Improper fractions are necessarily greater than one.

8.4 Fractions With Like Denominators

I claimed at the start of this chapter that multiplication and division of fractions is easy, and it is. I also claimed that addition and subtraction could be somewhat painful. However, when the fractions involved have the same denominator, addition and subtraction are as easy as multiplication and division. Here's the rule: *if both fractions have the same denominator, add or subtract the numerators and leave the denominator untouched*. For example, what's 2/5+2/5? The denominators are the same, so we add the numerators,

$$\frac{2}{5} + \frac{2}{5} = \frac{4}{5}$$

because $2+2=4$. Likewise with subtraction,

$$\frac{7}{9} - \frac{5}{9} = \frac{2}{9}$$

because $7-5=2$. In both cases, the denominator is unchanged. To see why is straightforward with the number line. Here's 2/5+2/5,

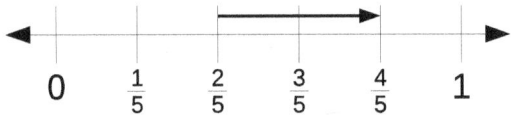

As with integers, we begin at 2/5 then add another 2/5, meaning move two pieces as the interval from 0 to 1 has been divided into five pieces.

The rule remains the same even if the result is an improper fraction, so $3/5 + 4/5 = 7/5$, which is quite reasonable. If we begin at 3/5 and count four more intervals of 1/5, we'll end up at 7/5, a number greater than 1 (= 5/5). There is nothing more to say; addition and subtraction of two fractions with the same denominator are no more challenging than addition and subtraction of two integers. Of course, we should remember to reduce the answer to lowest terms, if necessary.

8.5 Fractions With Unlike Denominators

It's a subjective statement, naturally, but many might agree that "The Good, the Bad, and the Ugly" was one of Clint Eastwood's best films. I haven't seen enough of his work to know, but I think it's a good film. Here, the good is that addition and subtraction of fractions with the same denominator are easy. The bad is that addition and subtraction of fractions with different denominators are not easy. The ugly is what we need to do to add or subtract such fractions.

How do we add 2/3 + 1/4? We can't add just the numerators because each fraction deals with different-sized pieces. The first has partitioned 0 to 1 into three pieces and the second into four pieces. The trick is to find a new name for each fraction by multiplying it by an alias for 1. The new name for each fraction is such that both end up with the same denominator. Once we have that, the good has returned: we add or subtract the numerators as before. One problem: *how* do we find the proper aliases for 1?

To answer that question, we need to introduce a new concept. We need to find the *least common multiple* (LCM) of the two denominators. Unlike "greatest common divisor," the phrase "least common multiple" makes more sense to us. We know the multiples of a number; we get them when we multiply the number by 1, then 2, then 3, and so on. So, the phrase implies that we should make a list of the multiples of each denominator and find the *smallest* multiple they have in common. That will help us find the proper aliases of 1.

Let's do it and see where we land. For the problem above, we need multiples of 3 and 4,

$$\begin{array}{ll} \text{Multiples of 3:} & 3, 6, 9, \mathbf{12}, 15 \\ \text{Multiples of 4:} & 4, 8, \mathbf{12}, 16, 20 \end{array}$$

I marked the smallest multiple in common, 12. Notice, in the list of multiples of 3, 12 is the fourth item because 3 × 4 = 12. Also, for the multiples of 4, 12 is the third item, 4 × 3 = 12. The aliases of 1 that we need are 4/4 and 3/3. Here's how we use them. First, multiply 2/3 by 4/4,

$$\frac{2}{3} \times \frac{4}{4} = \frac{8}{12}$$

Then, multiply 1/4 by 3/3,

$$\frac{1}{4} \times \frac{3}{3} = \frac{3}{12}$$

We now know that 2/3 = 8/12 and 1/4 = 3/12, and both fractions have the same denominator, so we replace 2/3 by 8/12 and 1/4 by 3/12, then add,

$$\frac{2}{3} + \frac{1}{4} = \frac{8}{12} + \frac{3}{12} = \frac{11}{12}$$

Let's try another example. Let's find 11/12 − 3/4. We begin by listing multiples of 12 and 4,

Multiples of 12: **12**, 24, 36
Multiples of 4: 4, 8, **12**, 16, 20

In this case, 12 is itself a multiple of 4, so we need not change 11/12 at all. If we did, we'd multiply it by 1/1 since 12 is the first multiple of 12. For the denominator of 4, we multiply by 3/3 to get 3/4 × 3/3 = 9/12. We finish the problem by straightforward subtraction of the numerators,

$$\frac{11}{12} - \frac{3}{4} = \frac{11}{12} - \frac{9}{12} = \frac{2}{12} = \frac{1}{6}$$

Notice, the final step is to reduce 2/12 by dividing the numerator and denominator by 2.

8.5.1 The Lazy

Let's not be the good, the bad, or the ugly. Let's be the lazy. Finding the least common multiple for each problem gets old quickly. There is another way that is often faster. Here's the trick: ignore the least common multiple. Instead, multiply the first fraction by the alias for 1 made from the denominator of the second, and the second by the alias using the denominator of the first. Let's try this with 11/12 − 3/4.

First, we multiply 11/12 by 4/4 because the denominator of the second number is 4. This gives us,

$$\frac{11}{12} \times \frac{4}{4} = \frac{44}{48}$$

8.5. FRACTIONS WITH UNLIKE DENOMINATORS

Second, we multiply 3/4 by 12/12 because the denominator of the first number is 12,

$$\frac{3}{4} \times \frac{12}{12} = \frac{36}{48}$$

Do you see what happened? In both cases, we end up with a denominator of 48. We can now subtract the numerators,

$$\frac{11}{12} - \frac{3}{4} = \frac{44}{48} - \frac{36}{48} = \frac{8}{48}$$

Finally, we reduce 8/48 using the trick above,

$$\frac{8}{48} = \frac{4}{24} = \frac{2}{12} = \frac{1}{6}$$

which is precisely what we found before.

Which approach to use is up to you. In general, I find it easier to multiply as above instead of finding the least common multiple. However, with practice, you'll start to see the LCM with no effort for many problems, but at first, you might find it easier to use the lazy approach. Either way, don't forget to reduce your final answer.

8.5.2 Section Summary

In this section, we learned two approaches to adding and subtracting fractions that have unlike denominators. The "correct" way first finds the least common multiple of the denominators of the fractions. Then, each fraction is multiplied by an alias for 1 made from whatever the denominator was multiplied by to get the least common multiple. Once the fractions have the same denominator, the addition or subtraction of the numerators finishes the calculation.

The "lazy" way of adding or subtracting fractions with unlike denominators multiplies the first by the alias of 1 using the denominator of the second, and vice versa for the second fraction, then completes the problem using the newly found names for the two fractions that now have the same denominator. In all cases, the final answer must be reduced to lowest terms.

We now know how to add, subtract, multiply, and divide any two positive fractions. Earlier, I referred to fractions as a ratio between two integers. Let's explore that idea in a bit more depth.

8.6 Ratios and Proportions

Let's explore yet one more way to think of a fraction, as a *ratio*. A ratio compares two numbers.

Ratios often appear in recipes, though implicitly. For example, if making rice on a stovetop, the ratio between the amount of water and the amount of rice is "2 to 1," meaning for every 2 cups of water, add 1 cup of rice. We can write the ratio as 2/1, though ratios are often written as 2 : 1 instead to clarify that we're thinking of a ratio.

Closely related to the idea of a ratio is that of a *proportion*. Proportions are helpful when we need to scale things, like a recipe. In this section, we'll briefly explore ratios and proportions.

8.6.1 Ratios

Ratios show the relationship between two numbers. The association between ratios and fractions is easy to see. Let's return to the rice recipe of 2 cups of water for every cup of rice, a 2 : 1 ratio. If we think of the entire pot, rice and water combined, as 1, we can use the ratio to understand what fraction is filled with rice and what fraction is filled with water. To do this, we transform the ratio into two fractions. The first fraction is the first number of the ratio with a denominator made by adding the two numbers together,

$$2:1 \rightarrow \frac{2}{2+1} = \frac{2}{3}$$

The second fraction is made the same way, but uses the second number,

$$2:1 \rightarrow \frac{1}{2+1} = \frac{1}{3}$$

The first fraction is how much of the pot is filled with water. The second is how much is filled with rice. So, 2/3 of the pot is water, and 1/3 of the pot is rice. Take care when assigning meaning to the parts of a ratio; we used 2 : 1 for water to rice. I could just as easily have used 1 : 2 for rice to water. Nothing would change other than the label on what each fraction represents. Notice also that,

$$\frac{2}{3} + \frac{1}{3} = \frac{3}{3} = 1$$

I.e., there are only two things in the pot, water and rice, so the respective fractions of each must add to 1, the entire pot.

8.6. RATIOS AND PROPORTIONS

Let's go the other way and convert fractions of a whole into a ratio. We have a box filled with red and blue marbles. There are 17 marbles total, 11 red and 6 blue. This means that 11/17 of the box is red marbles, and 6/17 of the box is blue.

To make the ratio, since the denominators are the same, we need only consider the numerators, 11 and 6. No calculation is necessary; we write 11 : 6 for the ratio between the red and blue marbles. Note, the ratio between the two fractions is really a division problem: $\frac{11}{17} \div \frac{6}{17}$. This makes sense as we said fractions are ratios and fractions are division problems as well. Let's work the problem explicitly,

$$\frac{11}{17} \div \frac{6}{17} = \frac{11}{17} \times \frac{17}{6}$$

$$= \frac{11 \times 17}{17 \times 6}$$

$$= \frac{11}{6} \times \frac{17}{17}$$

$$= \frac{11}{6} \times 1$$

$$= \frac{11}{6}$$

There are many steps here; let's cover them one by one. The first line rewrites the division as a multiplication, the usual way to divide two fractions. The second line shows what the multiplication does, the numerators are multiplied, and the denominators are multiplied.

The third line might seem like a sleight of hand trick, but it's perfectly valid. I'm rearranging the multiplication. If you multiply 11/6 by 17/17, you will do what the second line does.

The final two lines acknowledge that 17/17 is an alias of 1 and that something multiplied by 1 gives you that something back. Notice, the 17/17 trick only works because the two fractions have the same denominator.

What if the fractions do not have the same denominator? Now we have a box filled with red, green, and blue marbles, and we're told that the fraction of the box that is red marbles is 2/5 while the fraction that is blue marbles is 3/7. What is the ratio of red marbles to blue marbles?

In this case, we can't just use the numerators as they are because the denominators are not the same, we must first rewrite the fractions to use the same denominator by using the trick above that allowed us to add and subtract fractions with unlike denominators. Therefore,

$$\frac{2}{5} = \frac{14}{35} \text{ and } \frac{3}{7} = \frac{15}{35}$$

where the first fraction is multiplied by 7/7 and the second by 5/5. Now we can write the ratio using just the numerators: 14 : 15 for the ratio of red to blue marbles.

As an assignment for the reader: what fraction of the box is green marbles? I'll share the answer in the Section Summary below.

Ratios suffer from the same labeling problem as fractions; they must be reduced to lowest terms. Thankfully, reducing a ratio to lowest terms is the same as reducing a fraction to lowest terms. For example, suppose we calculated that the ratio between brown-eyed and blue-eyed people in a particular town is 42 : 18. This is a perfectly valid ratio, but it isn't as simple as it could be. The trick to reducing fractions applies here, divide both numbers first by 2, then 3, then 5, etc.,

$$42 : 18 = 21 : 9 = 7 : 3$$

The first step divided by 2 and the second by 3. So, the better answer to give is that there is a 7 : 3 ratio between brown-eyed and blue-eyed people in the town.

Ratios are most valuable when combined with proportions. Let's see how now.

8.6.2 Proportions

We've been using proportions throughout this chapter, but I haven't named them as such. More deception on the part of fractions!

Every time we change the label of a fraction by multiplying by an alias for 1, we use a proportion. Specifically, a proportion is an equation like this,

$$\frac{2}{5} = \frac{6}{15}$$

where we get 6/15 by multiplying 2/5 by 3/3. In general, though, the values in the proportion need not be integers, though we'll only work with integer proportions here. Let's return to the problem of

8.6. RATIOS AND PROPORTIONS

cooking rice. We know the ratio of rice to water is 1 : 2, that we need 1 cup of rice for every 2 cups of water. Now suppose, reasonably, that one cup of cooked rice is enough for one person. We're hosting a dinner party, and we have six guests coming. How much rice do we need? We need $6 \times 1 = 6$ cups. Great. Now, how much water do we need to cook the rice? Here's where the proportion comes in. What we are asking is: what value for x in this equation makes the equation true,

$$\frac{1}{2} = \frac{6}{x}$$

If we find an x that makes the equation true, meaning the fractions on either side of the equals sign are the same, we will know how much water we need.

There are ways to solve for x using the equation above and some algebra, but we don't need to go that far for a proportion. Consider the numerators. We multiply the 1 on the left by 6 to get the 6 on the right. We also know that the two fractions will be equal, meaning the first is multiplied by some alias of 1. If we multiplied the numerator by 6 to change 1 to a 6, then we need to multiply the denominator by 6 as well; this is the same as multiplying 1/2 by 6/6. Therefore, $x = 2 \times 6 = 12$, we need 12 cups of water to 6 cups of rice.

Let's work one more example, one that shows us we need not even write the proportion equation to scale a recipe. For this example, we'll use a recipe archeologists reverse-engineered from centuries-old bread discovered at a Viking campsite. Here's the recipe,

- 5 oz barley flour
- 2 oz wholemeal flour
- 2 tsp crushed flax seeds
- 1/2 cup water
- 2 tsp butter or lard
- salt to taste
- form into flat cakes about 1/4 inch thick
- bake in a 300 F oven for 10 to 13 minutes

I've made this bread several times. It's definitely filling. Let's say this recipe will feed four people. The bread is a hit, so your friend has asked you to bring some to her upcoming party. She tells you there will be 32 guests. How do you scale the recipe?

Your intuition, which is likely correct in this case, tells you that you need $32 \div 4 = 8$ times as much bread as the recipe makes if the recipe feeds four people, and 32 will be at the party. So, you figure, all you need do is multiply the amount of each ingredient by 8. Your intuition is correct. The ingredients are in the following ratio,

$$5 : 2 : 2 : \frac{1}{2} : 2$$

Here, I'm using the ":" notation to cover the relationship between all the ingredients at once; to give the ratio between the barley flour, wholemeal flour, flax seeds, water, and butter. To keep all the ingredients in the same ratio, we need to multiply each number in the ratio by the same value, 8 in this case, as we need to increase the recipe by a factor of 8. This is equivalent to multiplying by 8/8 for every ratio in the recipe. The result is,

$$40 : 16 : 16 : 4 : 16$$

meaning we need 40 oz barley flour, 16 oz wholemeal flour, 16 tsp crushed flax seeds, 4 cups of water, and 16 tsp of butter. Remember, $8 \times 1/2 = 8/2 = 8 \div 2 = 4$. In practice, you would naturally just multiply every ingredient by eight, but when you do that, you are scaling while keeping everything in proportion.

8.6.3 Section Summary

In this section, we briefly explored ratios and proportions, two topics related to fractions. First, we learned that ratios are comparisons between numbers, how one relates to another in terms of amount. Then, we learned how to turn two fractions representing parts of a whole, like marbles in a box, into a ratio.

Next, we examined proportions to understand that they help scale ratios. We realized that we've been using proportions all along every time we multiply a fraction by an alias of 1. We used the proportions of an ancient recipe to scale the recipe for a larger group of people.

Finally, I slipped a challenge problem into the section. Let's solve it now. We know that the fractions of red, green, and blue marbles in the box must add to 1, the entire box. So,

$$\frac{2}{5} + \frac{3}{7} + g = 1$$

8.7. NEGATIVE FRACTIONS

Here, g is the fraction of green marbles, what we want to find. To find it, we need to add 2/5 and 3/7, then subtract that total from 1. Adding gives us,

$$\frac{2}{5} + \frac{3}{7} = \frac{14}{35} + \frac{15}{35} = \frac{29}{35}$$

and subtracting gives,

$$1 - \frac{29}{35} = \frac{35}{35} - \frac{29}{35} = \frac{6}{35}$$

Therefore, 6/35, six thirty-fifths, of the box is green marbles.

8.7 Negative Fractions

Fractions act just like integers in terms of arithmetic operations and signs. Therefore, when multiplying and dividing fractions, these rules still apply,

- If *either* of the numbers is negative, the answer is *negative*.
- If *both* of the numbers are negative, the answer is *positive*.

Great. Now, what about addition and subtraction? Again, nothing changes. If we subtract a larger fraction from a smaller one, we still get a negative answer. Likewise, if we add a large positive fraction to a smaller negative fraction, we get a positive answer.

Perhaps the simplest way to see that fractions operate as integers for addition and subtraction is when the fractions involved have like denominators. We know we can always rename the fractions to have like denominators, so we need only consider that case. Also, if the fractions have like denominators, then addition and subtraction involve only the numerators, which are integers, so nothing changes in terms of signs. Let's show this with a few examples. Let's find $1/3 - 7/8$.

Using our trick from above, we multiply 1/3 by 8/8 and 7/8 by 3/3. This gives us,

$$\frac{8}{24} - \frac{21}{24} = \frac{8-21}{24} = \frac{-13}{24} = -\frac{13}{24}$$

because $8 - 21 = -13$. Notice, the negative sign on a fraction can go on the numerator or before the fraction. Either placement is correct. Putting the negative sign on the denominator is just plain weird, but if you do, multiply by $-1/-1$, a perfectly valid alias of 1, and you'll move the negative from the denominator to the numerator. Convince yourself that I'm not lying.

Let's try another. Let's find $-2/5 + {}^-7/9$. We want to add a negative value to a number that is already negative. Thinking of the number line, we are left of zero, at $-2/5$, and we are adding a negative, so we expect to move further left, away from zero.

We need like denominators. Nothing changes in that regard, we'll multiply $-2/5$ by $9/9$ and $-7/9$ by $5/5$. This gives us, $-18/45$ and $-35/45$. Now we add,

$$\frac{-18}{45} + \frac{-35}{45} = \frac{-18 + {}^-35}{45} = -\frac{53}{45}$$

The result is an improper fraction, here negative, and therefore greater than negative one, i.e., further from zero.

In general, to compare the size of two fractions, we need to convert them to the same denominator first. However, once we do that, we only need to consider the numerators to see the relationship.

8.7.1 Section Summary

Negative fractions exist and follow the same rules for multiplication and division as integers. We shouldn't be surprised by this as integers are a particular class of fraction, fractions with a denominator of 1.

This section also showed us that addition and subtraction problems involving negative fractions work like those involving negative integers. This fact is easy to see once the fractions are converted to like denominators.

8.8 Mixed Numbers and Improper Fractions (Oh, My!)

The title of this section sounds a bit like a 1950s high school dance chaperone's nightmare. I assure you, no funny business here. Okay, there is, but it isn't all that bad.

We already know that improper fractions are fractions where the numerator is larger than the denominator. We also know that improper fractions are necessarily bigger than one; we saw this with the number line and 3/2.

In this section, we learn a new notation, that of *mixed numbers*, and see how this new notation is actually a flight of fancy; in real-

8.8. MIXED NUMBERS AND IMPROPER FRACTIONS (OH, MY!)

ity, there are only improper fractions. Some math teachers might balk at my previous statement, but I stand by it. As we'll see, what we call a mixed number is, in reality, an addition problem with a missing symbol, nothing more. I'll introduce mixed numbers, then destroy them by showing that they are nothing more than sneaky improper fractions in disguise. Tradecraft has nothing on mathematics. However, in the process, we'll make an important discovery, one related to the division problems we solved in Chapter 6.

8.8.1 Mixed Numbers

Consider this number line,

What label should we give the question mark?

If we are thinking of mixed numbers, we use the label $1\frac{3}{4}$. Mixed numbers are mixed because they combine an integer (1) with a fraction (3/4). We read $1\frac{3}{4}$ as "one and three fourths." The "and" is a clue that we'll return to later.

In a way, it makes sense to use the label $1\frac{3}{4}$. From zero, we've moved to 1, but then we moved some distance past 1. How far? There are four pieces between 1 and 2 on the number line, so each piece is 1/4. The question mark is 3 of those 1/4 beyond 1, so that's the origin of 3/4 in $1\frac{3}{4}$.

What about this number line?

Give this one a try. I'll wait.

There are five divisions between 0 and 1. Likewise, there are five between 1 and 2, and 2 and 3. The question mark is between 2 and 3. How far? It's labeling the second piece, 2 out of the 5 pieces, so we write that the question mark is at $2\frac{2}{5}$, two and two fifths.

Mixed numbers are often used in the trades, so it is important to be familiar with them. The pain of mixed numbers comes from doing arithmetic with them. For example, lets add the two numbers we illustrated above, let's find $1\frac{3}{4} + 2\frac{2}{5}$. The trick is to add the integers, add the fractions, and then decide if we need to do anything more. So, we write,

$$1\frac{3}{4} + 2\frac{2}{5} = (1+2) + \left(\frac{3}{4} + \frac{2}{5}\right)$$

$$= 3 + \left(\frac{3}{4} \times \frac{5}{5} + \frac{2}{5} \times \frac{4}{4}\right)$$

$$= 3 + \left(\frac{15}{20} + \frac{8}{20}\right)$$

$$= 3 + \frac{23}{20}$$

Notice, I'm using parentheses to group the integers and the fractions while adding them, and just like with integers, we multiply before we add. So we end up with an integer, 3, and an improper fraction, 23/20, which is already in lowest terms.

Here's where an extra step comes in. The improper fraction is bigger than one, so we need to take one out of it, add it to the 3, and change the fraction into what's left over. Consider,

$$\frac{23}{20} = \frac{20}{20} + \frac{3}{20} = 1 + \frac{3}{20}$$

Therefore, we replace 23/20 with 1 + 3/20,

$$1\frac{3}{4} + 2\frac{2}{5} = 3 + \frac{23}{20}$$

$$= 3 + 1 + \frac{3}{20}$$

$$= 4 + \frac{3}{20}$$

$$= 4\frac{3}{20}$$

The final step above is another clue, $4 + 3/20 \to 4\frac{3}{20}$. Mixed number notation is nothing more than an addition problem with the "+" removed. Also, the process above is quite messy and ugly. But mixed numbers are a thing, and you'll run across them from time to time.

When I do, I tend to change the mixed numbers into improper fractions first. The next section shows us how. If you work in the trades where mixed numbers are most often used, you might choose to invest in a calculator that supports mixed number operations.

8.8.2 Mixed Numbers Are Improper Fractions in Disguise

Changing mixed numbers into improper fractions is straightforward, and the process will introduce us to an important fact. First, I'll demonstrate what's involved in converting a mixed number to an improper fraction; then, I'll show you a simple shortcut. After that, for completeness, we'll practice turning improper fractions into mixed numbers and discover a crucial consequence, one that completes our coverage of long division.

Mixed To Improper

Consider the mixed number $4\frac{7}{8}$, four and seven eighths. How should we transform this abomination into a "proper" improper fraction? First, contemplate what the mixed number is saying about itself,

$$4\frac{7}{8} = 4 + \frac{7}{8}$$

meaning the mixed number refers to the number that is 7/8 of the way from 4 to 5. How many eighths are there between 0 and 1? Eight. Likewise, there are eight eighths between any two integers. How many eighths, then, from 0 to 4? The answer is 4 × 8 = 32, eight added to itself four times. Therefore, we now know that,

$$4\frac{7}{8} = 4 + \frac{7}{8} = \frac{32}{8} + \frac{7}{8} = \frac{39}{8}$$

Stare at the above for a bit until it makes sense to you. The final step adds 32 and 7 since the denominators are, by design, the same. Recall, 32/8 = 32 ÷ 8 = 4.

You can use the above to turn a mixed number into an equivalent improper fraction, but there's a faster way. Try this,

$$4\frac{7}{8}$$

Follow the arrows from bottom to top. The first arrow is marked with "×", so multiply the 8 and the 4 to get 32. The second arrow has "+", so add 32 and 7 to get 39. The improper fraction is then 39/8, just as we found above.

Try this one, $9\frac{3}{7}$. First, $7 \times 9 = 63$ and $63 + 3 = 66$, so,

$$9\frac{3}{7} = \frac{66}{7}$$

Who doesn't love a shortcut?

Improperly Mixed

If, for some sad reason, you need to transform an improper fraction into a mixed number, the best way is to do what the fraction says: divide. For example, to transform 66/7 into a mixed number do,

$$\begin{array}{r} 9 \\ 7\overline{)66} \\ \underline{63} \\ 3 \end{array}$$

Here we learn something new, something quite important. In Chapter 6, we would write $66 \div 7 = 9r3$. We did that because we didn't yet know about fractions. The *true* answer is to take the remainder and write it as the numerator of a fraction with the divisor as the denominator. Therefore, the true answer is,

$$66 \div 7 = 9\frac{3}{7}$$

which is precisely the mixed number we found previously.

Let me repeat it: *when dividing, the remainder, if any, is written as the numerator of a fraction with the divisor as the denominator.*

For example, in Chapter 6 we solved $1439 \div 3$,

$$\begin{array}{r} 479 \\ 3\overline{)1439} \\ \underline{12} \\ 23 \\ \underline{21} \\ 29 \\ \underline{27} \\ 2 \end{array}$$

We now know that the answer is really $479\frac{2}{3}$, four hundred seventy-nine and two-thirds. Henceforth, when doing long division, you can write the remainder as a fraction of the divisor. In Chapter 10 we learn about decimal numbers, which will give us yet another way to write such quotients.

8.8.3 Section Summary

Mixed numbers were on the menu for this section. We were somewhat terse in our coverage, but we hit the salient points: mixed numbers are really compact addition problems, and mixed numbers and improper fractions go hand-in-hand. We saw an example of addition with mixed numbers, subtraction is much the same, but I recommend using improper fractions for multiplication and division. We learned the vital lesson that the remainder of a division problem is the numerator of a fractional part of the quotient with the divisor the denominator. Finally, we reviewed processes for turning mixed numbers into improper fractions and vice versa.

8.9 Chapter Summary

This chapter introduced a myriad of new concepts. First, we focused on the many disguises fractions wear. We learned that fractions are parts of a whole and exist on the number line between the integers. We also learned that mathematicians consider fractions, specifically rational numbers, as a class of numbers, including integers as a special case.

To work with fractions requires knowing how to reduce them to their lowest terms, to the unique base name that identifies a fraction and separates it from the infinite number of alias found by multiplying by an alias for 1. So, we learned that next, how to divide numerator and denominator by the greatest common divisor, or by following the trick of quick division by increasing primes. Along the way, we learned that multiplying two fractions is straightforward multiplication of numerators and denominators. We then discovered that division of two fractions is multiplication by the reciprocal of the second.

Addition and subtraction of fractions came next. The simple case, when the two fractions have the same denominator, adds or subtracts the numerators. Adding and subtracting fractions with unlike numerators requires finding the least common multiple of the denominators, or at least multiplying the denominators and using that to tell us which aliases for 1 we need to multiply each fraction by to get a common denominator. With the common denominator, we add or subtract the numerators. As always, reduce your answers as you go.

Fractions can also be thought of as ratios between two integers. Ratios are related to proportions, so we considered both. We learned how to use a proportion to scale a recipe to accommodate a specific number of people.

As fractions can be positive or negative, we briefly contemplated arithmetic with negative fractions and quickly saw that it follows the same rules as positive and negative integers.

We concluded the chapter with a disparaging look at mixed numbers and learned that mixed numbers are addition problems in disguise and can be turned into improper fractions with ease. Along the way, we learned that all the remainders we were writing in Chapters 6 and 7 are actually the numerators of fractions where the divisor is the denominator.

This is a lengthy and challenging chapter. The bare mechanics of arithmetic with fractions is the minimum goal, but I encourage you to reread the chapter and fight with the concepts, if any are still fuzzy. Chapter 9, to which we now turn, introduces us to exponents and powers, thereby setting the stage for Chapter 10 on decimals.

8.10 Terms and Concepts

We introduced the following terms and concepts in this chapter.

alias for 1 Any fraction of the form n/n for any $n \neq 0$. Any fraction where the numerator and denominator are the same is an alias for 1. To give fractions the same denominator, multiply by a suitably chosen alias for 1. (not an official math term)

denominator The second number of a fraction (the bottom). The denominator specifies the number of equal-sized parts the whole has been partitioned into.

fraction A fraction is a part of a whole. A fraction is a ratio between two integers. A fraction is a rational number.

greatest common divisor The greatest common divisor (GCD) is the largest factor in common between the numerator and

8.10. TERMS AND CONCEPTS

denominator. Divide the numerator and denominator by this number to reduce the fraction to lowest terms.

half Half is the distance from 0 to the middle of the distance between 0 and 1 on the number line. Dividing a whole into two equal-sized parts gives two halves. The fraction 1/2.

improper fraction A fraction where the numerator is larger than the denominator.

least common multiple The smallest number that is a multiple of both the numerator and the denominator. Use the LCM to change fractions so they have the same denominator to allow addition or subtraction.

lowest terms The unique base name for a fraction where the numerator and denominator have no factors in common.

mixed number A number format that combines an integer part with a fractional part. Mixed numbers are addition problems in disguise.

multiplicative inverse The multiplicative inverse of a number is the number that, when multiplied by the first, gives 1 as the answer. For fractions and integers, the multiplicative inverse is the reciprocal of the first number.

numerator The first number of a fraction (the top). The numerator "numbers" the denominator. It tells how many denominator-sized pieces of the distance from 0 to 1 on the number line the fraction contains.

proportion An equation equating two fractions. Multiplying a fraction by an alias for 1 produces a new version that is in the same proportion.

quarter Divide the distance from 0 to 1 into four equal-sized parts. One of those parts is a quarter. There are four quarters to a whole. For US coinage, a quarter is 25 cents. The fraction 25/100 reduces to 1/4, showing that a quarter (25 cents) is 1/4 of a dollar (100 cents).

ratio A relationship between two numbers. Fractions are ratios between two integers, the numerator and the denominator.

rational number A fraction, p/q, for any two integers, p and q, where q is not zero, is a rational number. The integers p and q are in the ratio $p:q$. Integers are rational numbers, those where $q = 1$.

reciprocal The number found by flipping the numerator and denominator of a fraction. The reciprocal of an integer, p, is the fraction $1/p$.

reduce The process by which the lowest terms representation of a fraction is found. Dividing the numerator and denominator by the greatest common divisor between reduces a fraction to lowest terms.

unit fraction A fraction where the numerator is one. In mathematics, the word "unit" often refers to one in some way.

8.11 Exercises

Exercise 1

Solve the given problem. Remember to express your answer in lowest terms.

$$\frac{6}{7} \div \frac{9}{6} \qquad \frac{6}{4} \times \frac{5}{6} \qquad \frac{7}{6} \times \frac{1}{3} \qquad \frac{9}{9} \div \frac{6}{2}$$

$$\frac{4}{3} + \frac{1}{2} \qquad \frac{4}{8} \times \frac{8}{7} \qquad \frac{3}{9} \div \frac{7}{3} \qquad \frac{6}{5} + \frac{5}{9}$$

$$\frac{6}{7} \div \frac{7}{2} \qquad \frac{6}{7} \div \frac{8}{2} \qquad \frac{2}{5} \div \frac{6}{2} \qquad \frac{5}{2} \div \frac{1}{9}$$

$$\frac{7}{2} - \frac{2}{8} \qquad \frac{7}{5} - \frac{6}{4} \qquad \frac{6}{3} + \frac{5}{6} \qquad \frac{8}{8} \div \frac{2}{6}$$

$$\frac{6}{8} - \frac{4}{5} \qquad \frac{7}{9} \times \frac{2}{4} \qquad \frac{5}{8} + \frac{6}{2} \qquad \frac{3}{8} + \frac{8}{6}$$

$$\frac{9}{8} + \frac{8}{7} \qquad \frac{7}{7} - \frac{6}{8} \qquad \frac{6}{9} \div \frac{3}{6} \qquad \frac{2}{5} \times \frac{4}{9}$$

$$\frac{1}{8} + \frac{6}{2} \qquad \frac{8}{8} \div \frac{1}{6} \qquad \frac{2}{4} - \frac{8}{5} \qquad \frac{3}{9} + \frac{4}{7}$$

8.11. EXERCISES

$$\frac{5}{4} \div \frac{4}{7} \qquad \frac{5}{7} \div \frac{9}{8} \qquad \frac{4}{9} \times \frac{2}{8} \qquad \frac{6}{2} + \frac{2}{4}$$

$$\frac{4}{9} + \frac{3}{7} \qquad \frac{4}{2} + \frac{5}{5} \qquad \frac{3}{6} \div \frac{7}{2} \qquad \frac{1}{3} - \frac{4}{5}$$

$$\frac{8}{3} - \frac{7}{8} \qquad \frac{3}{6} \times \frac{6}{3} \qquad \frac{2}{6} \div \frac{6}{3} \qquad \frac{7}{6} + \frac{3}{5}$$

Think About It

Flip a coin. What possible outcomes can you have, ignoring the unlikely one of it landing on its side? Two, heads or tails. There are two possible outcomes, and each flip will randomly return one of them. Probability, the likelihood of something happening is a number between 0 and 1. Fractions are typically numbers between 0 and 1. Fractions are often a helpful way to represent the probability of something happening.

The probability of a coin flip coming up heads is 1 out of 2. There are two possible outcomes, and one of them is to land heads up. The other is tails up, also 1 out of 2. The phrase "1 out of 2" means a ratio of 1 to 2, which we can, as you might have already guessed, write as 1/2. Therefore, the probability of a coin flip landing heads up is 1/2. The probability of landing tails up is also 1/2 as there is only one way to get a tail out of the two possible outcomes. Note, 1/2 + 1/2 = 2/2 = 1. Probabilities add to 1 when all possible outcomes are considered.

Roll a 6-sided die. What is the probability of getting a 3? There are six possible outcomes, and one of them is a 3, so the probability of getting a 3 is 1 out of 6: 1/6. Each number on a die is equally likely, that's the entire point of the die, so the probability of getting a 6 is also 1/6, and the sum of the six possible outcomes, all with probability 1/6, is 1/6 × 6 = 6/6 = 1.

If you count the number of possible ways the desired outcome can happen and divide it by the number of possible outcomes, you get the probability of that outcome. For example, if I roll two 6-side dice and add them, what is the probability of each possible outcome? Is the probability of getting a sum of 2 the same as the probability of getting a sum of 7?

Calculate the probabilities for each possible outcome for the sum of two 6-sided dice. The smallest sum is 1 + 1 = 2, and the largest is 6 + 6 = 12, so there are eleven probabilities to find. How many possible outcomes are there? For the first die, there are six and the same for the second, so there are 6 × 6 = 36 possible outcomes from rolling two dice.

I suggest making a table with 6 rows and 6 columns to represent the possible outcomes for the individual dies. Then, fill in the table by adding the row and column values. Something like this,

	1	2	3	4	5	6
1						
2						
3	4					
4						
5				9		
6						12

I've already filled in a few of the positions for you.

Finally, starting with 2, count each time a 2 appears in the table, then 3, then 4, etc., up to 12. Those numbers are the number of times each possible outcome can happen. To get the probabilities, form fractions with those counts as the numerators and 36 as the denominator as that is the total number of possible outcomes. You'll need to reduce the fractions to lowest terms.

To maximize your chance of winning, should you place your money on rolling a 4 or a 7?

Chapter 9

Powers

We have one final type of number to explore, the decimals. However, before we that, we need to make a slight detour to investigate powers and exponents. After the conceptual challenges of fractions in Chapter 8, powers and exponents are a welcome respite.

Let's investigate the idea of powers, including learning what I mean by that word and its related term "exponent." Along the way, we'll learn that, once again, zero causes some trouble. We'll sort it before exploring the most common form of exponentiation, squaring, and the associated inverse operation, square roots.

Next, we fiddle around with negative exponents. Negative exponents set the stage for Chapter 10, decimals. A quick look at the rules for multiplying and dividing with exponents follows.

We conclude the chapter with a fun dive into number bases, which use powers. Exploring different number bases reinforces our notions of place notation. In addition, understanding number bases provides context and a foundation for Chapter 10.

So, sit back, relax with your favorite beverage at hand, and read on; it's all good.

9.1 Powers and Exponents

Chapter 5 taught us that multiplication is repeated addition. *Exponentiation*, meaning using *exponents* to raise an integer to a specified *power*, is repeated multiplication. I'll explain what each of these terms means as we go along. A few examples will suffice.

Conceptually, 3×4 means to add three to itself four times. What might we make of this?

$$3 \times 3 \times 3 \times 3 = 3^4 = 81$$

On the left, we have 3 multiplied by itself four times. We know how to do that, start on the left, multiply 3×3 to get 9, then multiply $9 \times 3 = 27$ and finally $27 \times 3 = 81$. Alternatively, we might notice that each pair of $3 \times 3 = 9$ and that $9 \times 9 = 81$. Either way, we can find the correct answer.

The notation in the middle, 3^4, is new. This is exponentiation, the exponent is 4, and we are raising 3 to the fourth power,

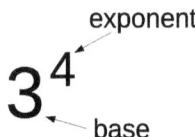

From the above, we see that raising an integer (or a fraction), the base, to an integer power, the exponent, is nothing more than multiplying the base by itself the exponent number of times. In general, any number can be raised to any other number, but in this book, we'll restrict ourselves to raising integers and fractions to integer powers. Note, however, that the exponent can be positive or negative.

Let's consider another example. Here's a small table of 2 raised to various powers. You might recognize some of these numbers, especially if you work with computers. Here's the table,

$$
\begin{aligned}
2^0 &= 1 \\
2^1 &= 2 \\
2^2 &= 4 \\
2^3 &= 8 \\
2^4 &= 16 \\
2^5 &= 32 \\
2^6 &= 64 \\
2^7 &= 128 \\
2^8 &= 256 \\
2^9 &= 512
\end{aligned}
$$

There are many things to observe in the table. The first is the confusing claim that $2^0 = 1$. We'll get to that in a bit. The second is that raising the base to the first power, an exponent of 1, gives you

9.1. POWERS AND EXPONENTS

the base back. This makes sense; if we have only the base, then, of course, we have the base alone as the answer.

A third observation is that each power is the base times the previous power. For example, $2^3 = 8$ and $2^4 = 2^3 \times 2 = 8 \times 2 = 16$. This also makes sense given how we calculate an expression where the same number is multiplied by itself repeatedly. Look above at how we calculated 3^4.

A fourth observation is that the number represented by a power snowballs as the exponent increases. This is the origin of the phrase *exponential growth*. For example, if we were to continue the table to an exponent of 32, the result is,

$$2^{32} = 4,294,967,296$$

This is a number greater than 4 billion. That's a rapid increase indeed. Many processes exhibit exponential growth, including the spread of a virus.

Misunderstanding exponential growth can be a critical mistake. For example, when I was in middle school, I borrowed a piece of paper from a friend. My friend, who knew about exponential growth, said that the amount of paper I owed him would double daily. On day zero, I owed him one sheet of paper; that's the $2^0 = 1$ I have yet to justify. On the next day, that one sheet doubled to 2 sheets. On day three, it was $2^2 = 4$ sheets, then 8, 16, 32, 64, etc.

I don't recall paying my friend back for the borrowed sheet of paper. A quick back-of-the-envelope calculation estimates that about 15,000 days have passed since then. How many sheets of paper do I owe? About $2^{15,000}$ sheets. That's a number with more than 4500 digits,

```
28179608796313976374286377853832223082416749129772963710783282778382317
93294568656027170971722368541300500148762830509570407079395897141508725
23073935114518202433817141315589046697665767753020703330561718642810157
43932289152388083699853148036051275473667695253212380344441315450388660
22013501130323430858329305553844388554400513754498442772182177128966753
63068720345515473434111699867542985885331741594726201346190830240692715
78112975456175929868418622669768839587283437535027206549575725916253189
38737134877529961701657660416458663546396286450553632434736911422560520
00072081219902252264258236028218613372215876496394341867072862629022156
55346553100944814868916568682518697494518571616628903566001018605941188
13679455546531878234387037341370764367875655580966233537465856288004554
30140169913540153961566294704894163494226738977413711366121112146042996
74235784883540126079790757274032586663484349209109819576179051096495220
10982042789342536193068324618351421696203455601605118404497018467666282
78420213281480510069379347007543973139554178175690773315927469361215139
95433957926598259428643472342342859072036357283455125014176157142919997
74537511019878238676975521858715895359821502314787149089305212445799600
30393194346889440433647348345868290410045628941650640315817691074244257
00875014126564652573885045749207567366536588266142970819560722413199837
39959574349583795051904587081024906887800893896551364917251585238845743
```

```
00750100902751828486806189811961720524504976040165319005498382580099971852
22872555140246658021274114527713438208485511921909732376032265141624589 11
05283200663509961444322011802982280199517308400644884871464258114559744500
95993844936566471047767974285413266855493150984091512278787539020882065 62
00893965077860471993210238274654894944955131195434242853848030257494862 59
26745047791730606995425888775571609281495953721339712877604791020000194 77
56249766479670986137076981224642743069725401099384031028670810757683522 82
16741189263735177183307060641786990352868151243180595299486637305753770 63
30058223974270942851702862462863025564046400969117960610444467223204786 97
09734836046571505598681429286372976726889778746623725623485244272247797 11
73030464647914414139566500486317063625500208906285521661046789938375490 91
77951472740881445817345223518733939048907047095904396482622164198141363 90
67723526075949807992319555177364587831440901887979371647118775640122112 69
14969173286729891857877863337439435712016824558497887684479410525891120 93
21714623819639305884042716787537989503386266503658086970090167114930782 28
67600084775797450043528335040523181300939047560092173656437964805986878 84
69097556168986643009081491211030654563468958955533468037622329954967569 24
18571358768898949962728257308339250406853786583881351167647082256831345 82
16527495376958238195629276140142912559084213140117194512209752564237149 30
45696984872342680046676856632775989569641163112112209709152721721827950 14
19161510202432408545231196334837237583982441706717084211494044627115285 05
52864783993722225100122068507428533164772669151799555291529447719732682 26
08680217636850523384079300468925004732873613241809253263851312834042402 75
85218868758335871701891694910725996305922129473237606780730600830901296 46
76038322426438467806570485299377626514669181680963962016304313021681669 69
65807537842952596401690070745066378443328338040451099819784763439422180 65
70802221756622288134849631040340416446743835264988563770278096937665079 97
49541038736604262724814821757258372470336571211357167468680200965241152 36
46708923282261877871087773232219881542898348263027454464445184244407000 33
94864980766527847507982666497421316255233645971997547872515731366992256 14
40432360473265250572606571269284353534676478375363478936179040075297456 77
38143449072131253334764766553664557330304609087121602080046206899547881 91
59704307313261037789301854306492695578195265021450463309721656154941131 51
19718460509357878230720821699669794678584993658379092808863297233433557 73
96239681834170033173019458984878205934376705558915485414696996442092694 44
31125131766865298990637518274963796773315169485343698526647970074976356 50
24494627271013468757521359567838089899510388135209838269214753262972848 93
93632953239135959856116943154062479347793351969778549744975948634474270 77
07960038638159313466532652012654426483204493696676837 2552526176229232138 1736
09626231926255343974851041449136902209592667970872917 6880055232106178255 0
19238512449404947081026614491855462404158037932313340926601173747329104 36
5718162108067861000003713852191284078882450691513817 08001509376
```

Clearly, it's essential to understand the idea of exponential growth. The number $2^{15,000}$ has no meaning in human terms. It's far, far, far greater than the estimated number of atoms in the entire visible universe. I sincerely hope my friend doesn't call me to collect on his debt.

To reiterate, raising an integer to a power is repeated multiplication. Let's take a look now at powers of fractions.

9.1.1 Powers of Fractions

What should we do with an expression like this?

$$\left(\frac{3}{4}\right)^5$$

9.1. POWERS AND EXPONENTS

First, let's understand what it even means. It means we want to raise the fraction 3/4 to the fifth power (notice how we speak of the powers, like fractions). Parentheses surround the fraction, then comes the exponent. This notation makes it clear that we want to raise the entire fraction to the fifth power.

If raising an integer base to a fifth power means multiplying the base by itself five times, we might think that the same holds true for a fraction, and we'd be correct. Let's work it out,

$$\left(\frac{3}{4}\right)\left(\frac{3}{4}\right)\left(\frac{3}{4}\right)\left(\frac{3}{4}\right)\left(\frac{3}{4}\right) = \frac{3^5}{4^5} = \frac{243}{1024}$$

The expression on the left demonstrates yet another way to specify multiplication, by writing numbers in parentheses side-by-side with no explicit × symbol between them.

We know how to multiply fractions together: multiply the numerators and denominators independently. Therefore, multiplying the 3 by itself five times is, as we just learned, 3^5. Likewise, for the denominator, we get 4^5. That explains the middle notation. It's still a fraction as both 3^5 and 4^5 are integers, 243 and 1024, respectively. As always, we need to reduce our fraction answers. In this case, 243/1024 is already in lowest terms.

For an integer base, the value of the power increases rapidly as the exponent increases. What happens as the exponent increases for a proper fraction (i.e., not an improper fraction > 1)? Let's explore using 1/2 as our sample fraction. First, what is one raised to any power? Well, it's 1 multiplied by itself that many times. However, $1 \times 1 = 1$, so we'll only ever get 1 as the answer no matter how many times we multiply. So, $1^n = 1$ for any integer we care to substitute for n. This simplifies our experiment, but the results will hold for any fraction where the numerator is less than the denominator, a necessary condition for a fraction to be less than 1.

We generated a table of powers of 2 above. For raising 1/2 to a power, the denominator will be just these values. Therefore, a table of 1/2 to the same powers gives us,

$$\left(\frac{1}{2}\right)^1 = \frac{1}{2} \qquad \left(\frac{1}{2}\right)^4 = \frac{1}{16}$$

$$\left(\frac{1}{2}\right)^2 = \frac{1}{4} \qquad \left(\frac{1}{2}\right)^5 = \frac{1}{32}$$

$$\left(\frac{1}{2}\right)^3 = \frac{1}{8} \qquad \left(\frac{1}{2}\right)^6 = \frac{1}{64}$$

As the exponent increases, the fractions get smaller and smaller. This trend continues. So, raising a proper fraction to ever-larger powers returns a number that gets closer and closer to zero. This is true even if the numerator is not 1. To be a proper fraction means the numerator is smaller than the denominator, therefore, as the exponent increases, the denominator will get larger faster than the numerator, and the fraction will still get closer and closer to zero.

9.1.2 What About Negative Bases?

There is no reason why we can't raise a negative integer (or fraction) to an integer power. The process involved remains the same, multiply the number by itself the exponent number of times. However, we have to be mindful of the final sign. Let's build a table of powers of -3,

$$\begin{aligned}
(-3)^0 &= 1 \\
(-3)^1 &= -3 \\
(-3)^2 &= 9 \\
(-3)^3 &= -27 \\
(-3)^4 &= 81 \\
(-3)^5 &= -243
\end{aligned}$$

Again, I'm using parentheses around the -3 to indicate that we intend to raise -3 to a power. A common mistake is to write -3^4, but that doesn't mean to raise -3 to the fourth power. Instead, it means raise 3 to the fourth power, then make the answer negative. In mathematics, we apply exponents before any other operation except using parentheses.

Look at the answers. What jumps out at you? I see the sign of the answer change as the exponents increase. I also see that negative numbers raised to the zeroth power are still one.

We should expect the signs to change. Each time we multiply two -3's together, we get a positive answer. Multiplying a positive number by a negative number gives us a negative number again. But multiplying that negative number by a negative number will again result in a positive number. So, the sign of the answer is toggled from positive to negative and back again as the exponent increases.

Also, look at the exponents producing negative answers. They are all odd numbers. Recall, an odd number is a number that ends in 1, 3, 5, 7, or 9. This is a general rule: *if the exponent is an odd number, and the base is a negative number, the answer is also negative.*

9.1.3 Section Summary

This section taught us what we mean by raising a base to a power using an exponent. We learned that for integer or fractional bases, the power is the base multiplied by itself the exponent number of times. We also learned that for integer bases, the answers get larger and larger as the exponent increases. For proper fractions, we saw that the opposite is true; as the exponent increases, the fraction gets smaller and smaller, closer and closer to zero. Finally, we learned that negative bases are allowed and that the sign of the result depends on the evenness or oddness of the exponent. Thus, a negative base raised to an odd exponent gives a negative result, while an even exponent gives a positive result.

Now, let's make good on a promise and address zero in the context of powers.

9.2 More Trouble With Zero

For any number n, be it an integer, a fraction, or otherwise, this is always true,

$$n^0 = 1$$

Why?

Earlier, I pointed out with the table of powers of 2 that each time the exponent was increased by one, the new result was 2 times the previous result. Now, think of the table in reverse order, where the exponents are getting smaller,

$$\begin{aligned} 2^4 &= 16 \\ 2^3 &= 8 \\ 2^2 &= 4 \\ 2^1 &= 2 \end{aligned}$$

Written this way, each row is the previous row divided by 2, and the exponent is one less. So, to continue the process, the next row should be 2^0, and it should be $2 \div 2 = 1$. Therefore, $2^0 = 1$. Also, regardless of the base, moving from an exponent of 1 to an exponent of 0 will result in the base divided by the base. But, we know that any number divided by itself is only an alias for 1, i.e., it gives 1 as the answer. Therefore, for any number at all, raising that number to the zeroth power gives 1 as the answer.

What about raising zero to a power? For example, what is 0^3? In this case, we do what the expression says, multiply 0 by itself three times. However, as $0 \times 0 = 0$, all we'll ever get as an answer is 0, no matter how many times we multiply zero by itself.

Finally, what is 0^0, zero raised to the zeroth power? By the first argument above, we might expect that we'll get 0/0, which we learned in Chapter 6 is indeterminate. In a sense, that is correct; it isn't wrong to think of 0^0 as indeterminate. However, in algebra and other branches of mathematics, $0^0 = 1$ is often used because it makes things nicer, not because it is necessarily correct. Most of the time, you are safe in assuming $0^0 = 1$.

9.3 Squares and Square Roots

Raising a number to the second power, like 4^2, is so common that it's worth investigating a bit, mainly because doing so leads to the inverse operation, the square root. First, though, why do we say raising a number to the second power is *squaring* it?

The answer comes from geometry. Here's a square where I marked the length of the side, 10 feet,

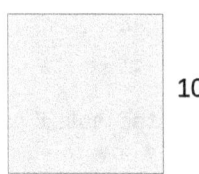

Let's imagine the square represents a small plot of land I want to cover with lavender to attract honey bees.

I need to know the area of the square because the bag of seeds says it covers an area of 25 square feet. The area of a square is the length of one side multiplied by itself, so the area of the small plot of land is,

$$10 \times 10 = 10^2 = 100 \text{ square feet}$$

Therefore, finding the area of a square involves raising the length of a side to the second power; hence, we are "squaring" it. I need $100 \div 25 = 4$ bags of seed to cover the plot.

9.3. SQUARES AND SQUARE ROOTS

Okay, now that we know the term's origin, what's special about squaring a number? Well, on the one hand, nothing, but on the other, plenty. Nothing because squaring is no different than raising to, say, the third power (called *cubing*) or any other power. But, plenty because many important things in mathematics and other disciplines involve squared quantities.

For us, squaring is a helpful transition into the inverse operation, the *square root*. I'm calling the square root the inverse of squaring because, in a sense, it undoes it. You'll understand what I mean as we go.

Consider this equation,

$$x^2 = 25$$

The equation asks, "what number, when squared, gives 25?" A few moment's thought will inform us that $x = 5$ is the answer because $5 \times 5 = 5^2 = 25$. Therefore, 5 is the square root of 25. *The square root of a number is the number that, when squared, gives you the original number.*

As squaring is very common in mathematics, dealing with square roots is also quite common. However, *finding* square roots by hand is not particularly simple. Also, most square roots are instances of yet another kind of number, one that we won't spend much time with.

Let's introduce some notation because this is math, and mathematicians love notation. How do we specify we want the square root of a number? Here's one way,

$$\sqrt{25} = 5$$

the $\sqrt{}$ symbol is called a *radical sign*. Read it as asking for the number that, when squared, gives you the number under the radical sign.

While every positive number has a square root, we'll only work with those that are *perfect squares*, meaning an integer squared, like 5 or 21, etc. For example, what is $\sqrt{2}$? There is an answer to that question, but the answer isn't an integer. We can see that it isn't an integer because the square root of 1 ($\sqrt{1}$) is 1, and the square root of 4 ($\sqrt{4}$) is 2. There are no other integers between 1 and 2. However, the square root of 2 ($\sqrt{2}$) *must* be between $\sqrt{1}$ and $\sqrt{4}$, which means it's between 1 and 2.

Not only that, $\sqrt{2}$ is a number that cannot be written as a fraction, i.e., $\sqrt{2}$ is not a rational number. Mathematicians call such numbers *irrational*. Who knew mathematicians could be so mean?

Given that we know the answer will be an integer, how do we find the square root? There is an algorithm akin to long division, but frankly, it isn't worth knowing in our current age. Just use a calculator.

However, before reaching for the calculator or running the calculator app on your phone, you can often quickly bracket the square root by trial multiplications. For example, what is $\sqrt{625}$? I'll tell you in advance that 625 is a perfect square, so there is an integer answer.

Take a guess, say 10. But, $10 \times 10 = 100$, so 10 is too low. Try 30. Here $30 \times 30 = 900$, which is too big. Halfway between 10 and 30 is 20, let's try that: $20 \times 20 = 400$. We're getting closer. What about halfway again between 20 and 30, which is 25: $25 \times 25 = 625$. Nailed it, $\sqrt{625} = 25$.

Other approaches to finding square roots require decimal arithmetic and iteration, where an initial guess is refined repeatedly until we get the answer to sufficient precision. Your best bet is to use a calculator except for simple square roots of perfect squares.

To find the square root of a fraction, find the square root of the numerator and denominator independently,

$$\sqrt{\frac{4}{9}} = \frac{\sqrt{4}}{\sqrt{9}} = \frac{2}{3}$$

9.3.1 Square Roots of Negative Numbers

The expression $\sqrt{1}$ is asking for the number that, when squared, gives 1 as the answer. We know this one: $1 \times 1 = 1$, therefore $\sqrt{1} = 1$.

Now, what is $\sqrt{-1}$? It's the number that, when multiplied by itself, gives -1 as the answer. Here we have a problem. Multiplying $-1 \times -1 = 1$ because the product of two negatives is always a positive number, as we learned in Chapter 5. There is no number that, when multiplied by itself, will give -1 as the answer.

Does that mean $\sqrt{-1}$ has no meaning? In a sense, yes, but in another sense, no. Mathematicians give a special name to $\sqrt{-1}$. They call it *i* for "imaginary." *Imaginary numbers*, also called *complex numbers*, are widely used in mathematics, science, and engineering. However, they are beyond what we need concern ourselves with here.

9.3.2 Section Summary

In this section, we learned what a square is and why it is called "a square." We then learned about the inverse operation, the square root. We learned that the square of an integer is called a perfect square and that we can sometimes find the square root of a perfect square by a few trial multiplications.

Additionally, we learned that while every positive number has a square root, some square roots represent a new kind of number, the irrational numbers, those that cannot be written as a fraction. Then, stretching our minds a bit, we learned that while negative numbers do not have an actual square root, it is possible to conceive of something like $\sqrt{-1}$ to create an imaginary number, one that is surprisingly valuable for more advanced mathematics.

So far, all our exponents have been positive integers. Let's mix things up now and contemplate negative exponents.

9.4 Negative Exponents

The thing to remember about negative exponents is this,

$$x^{-n} = \frac{1}{x^n}$$

Here, x is the base and n is an integer exponent. So, one way to look at negative exponents is that they are a shorthand for a unit fraction. A few examples will help clarify. Consider,

$$2^{-3} = \frac{1}{2^3} = \frac{1}{8}$$

and,

$$3^{-4} = \frac{1}{3^4} = \frac{1}{81}$$

In Chapter 8, we learned that the reciprocal of a fraction is the number found by flipping the numerator and denominator. Recall, also, that integers are fractions with a denominator of 1. Therefore, a negative power is the same as finding the reciprocal of the base, then applying the power. For example, 2^{-3} is identical to,

$$\left(2^{-1}\right)^3 = \left(\frac{1}{2}\right)^3 = \frac{1^3}{2^3} = \frac{1}{8}$$

Here, 2^{-1} indicates the reciprocal. This fact implies that raising a fraction to a negative power is the same as flipping the numerator and denominator before applying the exponent,

$$\left(\frac{3}{5}\right)^{-2} = \left(\frac{5}{3}\right)^{2} = \frac{5^2}{3^2} = \frac{25}{9}$$

Earlier, we made a small table of 1/2 raised to several different powers. We see now that this is no different than raising 2 to the negative version of those powers,

$$2^{-1} = \frac{1}{2} \qquad 2^{-4} = \frac{1}{16}$$
$$2^{-2} = \frac{1}{4} \qquad 2^{-5} = \frac{1}{32}$$
$$2^{-3} = \frac{1}{8} \qquad 2^{-6} = \frac{1}{64}$$

We'll make use of this observation in Chapter 10. Now, let's take a quick look at multiplying and dividing with powers of a common base.

9.5 Multiplication and Division with Powers

Multiplying and dividing with powers of the same base number to some exponent is straightforward. For example, consider,

$$(2^2)(2^3) = (4)(8) = 32 = 2^5$$

Notice, the product is the same as the base raised to the *sum* of the exponents, $2 + 3 = 5$. Consider another example,

$$(3^3)(3^4) = (27)(81) = 2187 = 3^7$$

Again, the answer is the base raised to the sum of the exponents. This is a general rule,

$$(x^n)(x^m) = x^{(n+m)}$$

for some base x and two integer exponents, n and m. Therefore, to find the product of a base to some power times the base to another power, add the exponents.

The exponents add even if they are negative. For example,

9.5. MULTIPLICATION AND DIVISION WITH POWERS

$$(2^{-2})(2^3) = 2^{(-2+3)} = 2^1 = 2$$

because $-2 + 3 = 3 - 2 = 1$.

Division is similar, but instead of adding, the exponents are subtracted, the second from the first. For example,

$$3^4 \div 3^3 = 3^{(4-3)} = 3^1 = 3$$

which is more often written as,

$$\frac{3^4}{3^3} = 3^{(4-3)} = 3^1 = 3$$

Notice, I'm using the slash or fraction form to indicate division. In general, we can write,

$$\frac{x^n}{x^m} = x^{(n-m)}$$

meaning the division of two powers of the same base, x, is the base raised to the difference between the powers, $n - m$.

Negative exponents work in division as well,

$$\frac{2^3}{2^5} = 2^{(3-5)} = 2^{-2} = \frac{1}{2^2} = \frac{1}{4}$$

and,

$$\frac{3^2}{3^{-3}} = 3^{(2-\,^-3)} = 3^{2+3} = 3^5 = 243$$

Notice, I dropped the parentheses around the exponent and wrote 3^{2+3}. The exponent of the base is the entire expression in small type, so, formally, parentheses are not needed.

One last question before we move on. What happens if we raise a power to a power? For example, what might this be?

$$(2^3)^2$$

If you think it means to raise 2^3 to the second power, you are correct,

$$(2^3)^2 = (2^3)(2^3) = (8)(8) = 64 = 2^6$$

But, notice the final expression, $2^6 = 64$. When you have a base raised to a power, then raise that expression to another power, the result is the same as raising the base to the product of the exponents. Using x as the base and n and m as stand-ins for the exponents, we can write,

$$(x^n)^m = x^{nm}$$

where the expression nm means to multiply n and m.

9.5.1 Section Summary

The key takeaways for the multiplication and division of the same base to powers is that when multiplying, *add* the exponents, and when dividing, *subtract* the exponents.

Let's conclude the chapter by having some fun with place notation. Doing so provides context for our exploration of decimal numbers in Chapter 10.

9.6 Number Bases

Back in Chapter 1, we discussed place notation as a way to represent numbers. We used a base of 10 because that's what most human cultures instinctively used. We said it then, but we'll demonstrate it now, that there is nothing special about using 10 as the base. If we had evolved to have 14 fingers on both hands, a distinct possibility as many early tetrapods had 7 digits per foot, we'd likely use 14 as the base for our number system.

As a review, here are the base 10 place names, but this time I'm adding something new, I'm adding the power of 10 that each place represents,

10^0	1	ones
10^1	10	tens
10^2	100	hundreds
10^3	1,000	thousands
10^4	10,000	ten thousands
10^5	100,000	hundred thousands
10^6	1,000,000	millions

Notice, because we use base 10, powers of 10 are especially simple to calculate: 10^n is a 1 followed by n zeros. We'll get to negative powers of 10 in the next chapter. Similarly, multiplying by any power of ten simply adds that many zeros to the number,

9.6. NUMBER BASES

$$123 \times 10 = 1230$$
$$2094 \times 100 = 209{,}400$$
$$88 \times 10{,}000 = 880{,}000$$

Earlier, we wrote numbers in expanded form, where the number is each digit value times the value of the place in the number. We can do the same here, but we'll use the power of 10 that each place represents instead. For example,

$$1234 = 1 \times 10^3 + 2 \times 10^2 + 3 \times 10^1 + 4 \times 10^0$$

Written this way, we can improve our understanding of place notation as each place is the base, 10, raised to a power. The values of the places are not arbitrary but follow a regular pattern; each place to the left is one exponent value greater than the place immediately to the right.

To write base 10 numbers, we need ten unique digit symbols, the digits 0 through 9. If the base of our number system is n, then we need n unique digit symbols.

Now, I'm going to replace base 10 with a different base, in this case, base 7. To write base 7 numbers, we need seven digit symbols, one for zero and six more for the numbers 1 through 6. Let's have some fun and invent our own symbols. There is no reason why we can't. Here's our set of symbols and the corresponding digit values using base 10 digits,

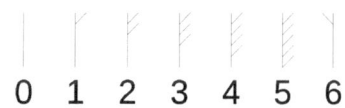
0 1 2 3 4 5 6

I intentionally made the symbols similar to tally marks so we could identify them quickly. They look a bit like ancient runes to me.

If we use base 7, then the place values are no longer based on 10 (literally) but instead use powers of 7,

$$\begin{array}{rr} 7^0 & 1 \\ 7^1 & 7 \\ 7^2 & 49 \\ 7^3 & 343 \\ 7^4 & 2{,}401 \\ 7^5 & 16{,}807 \\ 7^6 & 117{,}649 \end{array}$$

Here, I'm writing the value of the place with ordinary base 10 numbers. To use our new base 7 symbols, we'd write,

7^0 ┆

7^1 ┆|

7^2 ┆||

7^3 ┆|||

7^4 ┆||||

7^5 ┆|||||

7^6 ┆||||||

Our new base 7 symbols follow the same overall pattern as base 10. Multiplying by 7 adds a new zero symbol, here a vertical line.

The base 10 number 1234, when written in base 7, which I won't show how to determine here, is a four-digit number,

Writing 1234 in base 7 in expanded form using normal digits gives us,

$$1234 = 3 \times 7^3 + 4 \times 7^2 + 1 \times 7^1 + 2 \times 7^0$$

Notice that the form of the expression is the same as with base 10; if you replace "10" with "7."

As I said, there is nothing special about base 7 or base 10. So, are other number bases used, or is this just a fun diversion? Computers operate in base 2, called *binary*. Anyone involved with computers will eventually encounter base 2 numbers, and other numbers using bases that are powers of 2, like base 8 (*octal*) and base 16 (*hexadecimal*). We used binary numbers, indirectly, in the *Think About It* section of Chapter 5 when discussing ancient Egyptian multiplication.

If the base is 2, then the only digits we need are 0 and 1. The first nine powers of 2 beginning with 2^0 are 1, 2, 4, 8, 16, 32, 64, 128, 256. Here are a few binary numbers and their base 10 equivalents,

1010_2 10
101010_2 42
1100100_2 100
101100111_2 359

9.6. NUMBER BASES

I added a "2" subscript for the binary numbers to remind you that the number is in base 2. As an exercise, see if you can prove that the conversion to base 10 is correct. Converting a binary number to base 10 is straightforward. Simply add up the place values of every 1 digit.

Goofing around with different number bases is fun, but, seriously, what's the point? The point is to prepare us for Chapter 10 by learning that the value associated with each place is a specific power of the base, and that the exponents increase by one as we move from the right to the left.

But, more fundamentally, the point is to remind you that place notation is just that, a notational system we use to express a number; it is not the number itself. For example, there is truly no difference between

$$1234 \quad \text{and} \quad \text{⌐⌐⌐⌐}$$

Both representations refer to the same number, the same position on a number line. The second representation seems alien, but that's only an artifact of growing up using the first representation. The Apollo moon landings would have been just as successful if humanity had decided to use base 7 and our rune-like number symbols.

Whether or not the number we write in base 10 as 1234 actually *exists* is above our pay grade to discuss. Should you be so inclined, I recommend this article, https://www.welovephilosophy.com/2012/12/17/do-numbers-exist/. Enjoy, and keep some aspirin handy.

9.6.1 Section Summary

The purpose of this section is to present place notation as some base, usually ten, expressed as increasing powers beginning with an exponent of zero. A set of symbols, like the digits 0 through 9 for base 10, are necessary to express numbers in a particular base, but the choice of base itself is arbitrary in that there is no a priori preferred base.

Place notation exponents begin with 0 and increase by one for each new digit to the left. What do you think might happen if we were to add new places to the right of the first place instead of adding new places to the left? What exponents might we then need? Chapter 10 holds the answer.

9.7 Chapter Summary

This chapter introduced us to the idea of exponentiation, or powers, as repeated multiplication. We learned a new notation to indicate raising a number, the base, to a power by writing the exponent as a superscript on the base. We learned how to evaluate powers for integer and fractional bases, both positive and negative.

As with multiplication and division, zero again posed a bit of a problem. We learned that while 0^n has a definite meaning, 0^0 is troubling and can be viewed as indeterminate or given the value of 1 to make other math work nicely.

Raising a number to the second power, squaring, is particularly common in many disciplines, so we gave that process special attention, along with the inverse operation, the square root. Along the way, we discovered that trying to find the square root of a negative number introduces an entire can of worms, but a handy can of worms, all based around the labeling of $\sqrt{-1}$ as i, the imaginary number.

We pondered what might be meant by a negative exponent and learned that negative exponents imply the reciprocal operation followed by using the positive version of the exponent. With negative exponents in hand, we briefly considered the rules for multiplying and dividing the same base raised to different powers and learned that multiplication becomes addition of the exponents, while division becomes subtraction.

We concluded the chapter by experimenting with different number bases and learned that for any selected base, place notation uses powers of that base to define the value associated with that place. We ended the chapter by asking what it might mean to add new digits to place notation by moving to the right of the ones place instead of to the left.

We are now ready to tackle our final number format, the decimals. Put on your armor, draw your sword, and prepare for battle.

9.8 Terms and Concepts

We introduced the following terms and concepts in this chapter.

base   The number which is being raised to the power specified by the exponent.

binary   The base 2 number system. Computers operate in binary.

9.8. TERMS AND CONCEPTS

complex number Another name for an imaginary number.

cube Raising a number to the third power is cubing the number. The volume of a cube with sides of length a is a^3.

exponent The number of times the base is multiplied by itself. E.g. $2^3 = 2 \times 2 \times 2$ with base 2 and exponent 3.

exponential growth What happens when the amount of something increases by a multiple of its current value so that the amount of it is given by a base to some power. The spread of viruses often follows exponential growth.

exponentiation The act of raising a base to a specified power.

hexadecimal The base 16 number system, frequently used when working with computers. Hexadecimal numbers are typically written using the digits 0 through 9 and the letters A through F for the digits representing 10 through 15.

imaginary number The imaginary number i is defined to be the square root of -1.

irrational number A number that cannot be expressed as the ratio of two integers, i.e., as a fraction. For example, $\sqrt{2}$ is an irrational number because there are no integers, p and q, such that $p/q = \sqrt{2}$.

octal The base 8 number system sometimes used with computers.

perfect square A number is a perfect square if there is an integer that gives the number when multiplied by itself. E.g., 36 is a perfect square because $6 \times 6 = 36$.

power Raising a base to an exponent. For example, 4 to the third power is $4^3 = 4 \times 4 \times 4 = 64$.

radical sign The symbol used to indicate a square root. In the expression $\sqrt{3}$, the $\sqrt{}$ symbol is the radical sign.

square Raising a number to the second power, e.g., $3^2 = 9$ means 9 is 3 squared. The area of a square with length a is a^2.

square root The inverse operation to squaring. The square root of a number is the number that, when squared, gives the first number. E.g., the square root of 196 is 14 because $14^2 = 196$.

9.9 Exercises

Exercise 1

Write the answers to the following multiplication and division problems using the same base and the proper exponent. Naturally, you don't need to do every problem, just as many as you need to be comfortable with the process.

$7^5 \times 7^7$ $\qquad$ $7^8 \times 7^{-9}$ $\qquad$ $4^4 \times 4^6$ $\qquad$ $4^4 \div 4^5$

$3^{-3} \times 3^{-2}$ $\qquad$ $2^5 \times 2^{-7}$ $\qquad$ $9^2 \div 9^{-5}$ $\qquad$ $3^7 \div 3^{-3}$

$4^5 \times 4^9$ $\qquad$ $5^{-5} \div 5^5$ $\qquad$ $2^{-7} \div 2^5$ $\qquad$ $3^4 \div 3^{-4}$

$5^8 \times 5^{-9}$ $\qquad$ $8^{-9} \div 8^{-9}$ $\qquad$ $2^7 \div 2^5$ $\qquad$ $8^2 \div 8^9$

$9^8 \times 9^{-8}$ $\qquad$ $5^5 \times 5^{-2}$ $\qquad$ $7^2 \div 7^2$ $\qquad$ $8^8 \times 8^7$

$2^9 \div 2^{-5}$ $\qquad$ $3^6 \div 3^4$ $\qquad$ $5^5 \div 5^{-8}$ $\qquad$ $6^3 \div 6^{-3}$

$2^4 \div 2^5$ $\qquad$ $8^5 \times 8^{-2}$ $\qquad$ $7^2 \div 7^{-8}$ $\qquad$ $2^3 \times 2^3$

$6^5 \times 6^7$ $\qquad$ $7^7 \div 7^6$ $\qquad$ $2^8 \times 2^8$ $\qquad$ $3^5 \times 3^{-7}$

$7^7 \times 7^{-6}$ $\qquad$ $4^3 \times 4^9$ $\qquad$ $3^4 \times 3^5$ $\qquad$ $2^5 \times 2^8$

$5^2 \div 5^{-5}$ $\qquad$ $2^4 \div 2^8$ $\qquad$ $6^8 \div 6^9$ $\qquad$ $2^{-2} \div 2^{-3}$

Exercise 2

Find a number to substitute for x to make the following equations true. Recall, an equation is true if both sides of the equals sign have the same value.

9.9. EXERCISES

a) $x^2 = 169$
b) $\sqrt{x} = 10$
c) $\sqrt{\dfrac{9}{16}} = x$
d) $\sqrt{25} + x = \sqrt{36}$

Think About It

Base 3 uses the digits 0, 1, and 2. When adding, carry on 3, not 10. Complete the addition and multiplication tables for base 3, then solve the problems below. I'll start both tables for you,

+	0	1	2
0	0	1	2
1			10
2	2		

×	0	1	2
0	0	0	0
1	0	1	
2	0	2	

Add: $\begin{array}{r} 21012 \\ +\ \ 2212 \\ \hline \end{array}$ Multiply: $\begin{array}{r} 221 \\ \times\ \ 12 \\ \hline \end{array}$

What two holidays are programmers always confusing?

Chapter 10

Decimals

We discussed many kinds of numbers in this book. In this chapter, we encounter one last kind of number, the decimals. Decimal numbers are place notation's version of fractions. You'll see what I mean by that as the chapter progresses.

We start the chapter with an essay, "Concerning Decimals." The essay walks us through the essence of decimal numbers. After that, we begin learning how to add, subtract, multiply, and divide with decimal numbers. Along the way, we briefly explore comparing decimal numbers with each other to see which is larger.

Next, we examine converting fractions into decimals. Converting fractions to decimals teaches us some interesting things about numbers in general. At last, we are ready to dive into the division of decimal numbers, first by powers of ten, then by integers, and finally by other decimals. We conclude the chapter with a brief look at how computers manipulate decimal numbers.

10.1 Concerning Decimals

Chapter 9 asked the question: what might it mean for place notation to have digits to the right of the ones column? I'll answer that question now via example,

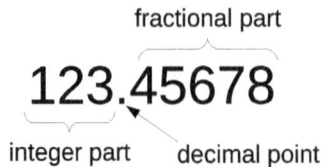

Above is a *decimal number*. I labeled its parts. Let's begin in the middle, with the *decimal point*. The decimal point is a marker telling us where the ones place is; immediately to the *left* of the decimal point, where the 3 is in this example.

To the left of the decimal point, we have the *integer part*. You'll often see this referred to as the *whole number* part. However, we know that numbers can be positive or negative, so we'll use the adult term: integer part.

So far, I haven't described anything new. Adding a decimal point after the ones column doesn't affect how we read the number. What *is* new is the *fractional part*, the numbers to the right of the decimal point. That's what makes this a decimal number.

A brief comment about notation is again necessary. Using a period as the decimal point (technically a *decimal separator*) is a North American convention along with the UK, Ireland, Australia, and China, among others. Much of the remainder of the world use a comma to separate the integer part from the fractional part. We'll remain North America-centric and use a period reserving commas for grouping digits of the integer part.

10.1.1 The Fractional Part

What should we make of the five digits after the decimal point in the example above? The name gives us a clue: fractional. In a decimal number, the portion to the right of the decimal point represents a fraction, and more than that, a fraction that is less than one.

I think, given our newfound expertise with powers, that the best way to understand what the fractional part represents is to continue with expanded notation. The example above is really the following number,

$$1 \times 10^2 + 2 \times 10^1 + 3 \times 10^0 + 4 \times 10^{-1} + 5 \times 10^{-2} + 6 \times 10^{-3} + 7 \times 10^{-4} + 8 \times 10^{-5}$$

We are familiar with the first three parts, those with positive exponents of 10. And, we now see that the fractional part of a decimal number is shorthand for a set of fractions, those formed by the

10.1. CONCERNING DECIMALS

negative exponents of 10. Read the exponents from left to right: 2, 1, 0, −1, −2, −3, −4, −5. Each place value to the right is the current value divided by 10.

A decimal number is, in its essence, shorthand for a mixed number. For example, while cumbersome, this representation denotes the exact same number as 123.45678,

$$123 + \frac{4}{10} + \frac{5}{100} + \frac{6}{1000} + \frac{7}{10,000} + \frac{8}{100,000}$$

Here, I've written each negative power of 10 as a fraction which is then multiplied by the digit value. Recall,

$$10^{-1} = \frac{1}{10}, \quad 10^{-2} = \frac{1}{100}, \quad 10^{-3} = \frac{1}{1000}, \quad 10^{-4} = \frac{1}{10,000}, \quad 10^{-5} = \frac{1}{100,000}$$

The fractional part of a decimal number tells us how many tenths, hundredths, thousandths, ten thousandths, etc., to add to the integer part. Another way to think of it is that the fractional part represents the fraction of the distance on the number line beyond the integer part towards the next higher integer. For our example, this means the fractional part represents how much further along the number line we need to go beyond 123 towards 124, the next integer.

How is this different from the mixed numbers I was so mean to in Chapter 8? The difference has to do with the denominators of the fractions; they are all powers of 10, and 10 is the base of our number system. This fact lets us write the fractions in a shorthand way, just as we write the integer part.

Let's switch to a simpler decimal number, 1.7. If we understand what 1.7 represents, we're well on our way to understanding any decimal number.

The integer part is 1. The fractional part is 0.7. Notice, I put a zero for the integer part. This helps in reading the number. Of course, it's perfectly okay to write .7 as the fractional part, but it's easy to miss the decimal point, so adding the 0 helps and does not change the meaning.

Using the example above as a guide, the fractional part represents,

$$0.7 = 7 \times 10^{-1} = 7 \times \left(\frac{1}{10}\right) = \frac{7}{10}$$

Pause here and let the above sink in. The first digit of the fractional part of a decimal number counts the tenths. For 1.7, we have seven of them. If we have 9 tenths and we add one more tenth, we have 1, like so,

$$\frac{9}{10} + \frac{1}{10} = 1 = 0.9 + 0.1$$

Notice, the left expression and the right expression are *exactly* the same. The left one uses the fractions we are familiar with. The right one uses our new decimal notation now that we know the first digit to the right of the decimal point counts tenths. Let's write the decimal notation part vertically,

$$\begin{array}{r} 1 \\ 0.9 \\ +0.1 \\ \hline 1.0 \end{array}$$

Interesting, adding 0.9 and 0.1 to get 1.0, meaning one and no tenths, looks a lot like adding 9 and 1 to get 10 if we ignore the decimal point. I even show a carry of one. Hmm. Keep this observation in the back of your mind.

Let's expend a bit more effort exploring what the fractional part of a decimal number means. Let's work with this number,

$$2.3149 = 2 \times 10^0 + 3 \times 10^{-1} + 1 \times 10^{-2} + 4 \times 10^{-3} + 9 \times 10^{-4}$$
$$= 2 + \frac{3}{10} + \frac{1}{100} + \frac{4}{1000} + \frac{9}{10,000}$$

I'm writing 2.3149 first in expanded notation using negative exponents of 10. The second version uses fractions instead. Both representations are valid.

What happens if we decide to add each of the four fractions of the second representation? Let's do it beginning with the tenths and hundredths,

$$\frac{3}{10} + \frac{1}{100} = \left(\frac{3}{10}\right)\left(\frac{10}{10}\right) + \frac{1}{100} = \frac{30}{100} + \frac{1}{100} = \frac{31}{100}$$

I need to multiply 3/10 by the 10/10 alias for 1 to get a common denominator of 100. Now, let's add in the next fraction in the expansion of 2.3149,

$$\frac{31}{100} + \frac{4}{1000} = \left(\frac{31}{100}\right)\left(\frac{10}{10}\right) + \frac{4}{1000} = \frac{310}{1000} + \frac{4}{1000} = \frac{314}{1000}$$

10.1. CONCERNING DECIMALS

and, adding the final fraction gives us,

$$\frac{314}{1000} + \frac{9}{10,000} = \left(\frac{314}{1000}\right)\left(\frac{10}{10}\right) + \frac{9}{10,000} = \frac{3140}{10,000} + \frac{9}{10,000} = \frac{3149}{10,000}$$

Our answer is the fractional part of 2.3149. So, the decimal number is really,

$$2.3149 = 2 + \frac{3149}{10,000}$$

This result isn't an accident. We can always write the fractional part of a decimal as a fraction with the denominator a power of 10. Which power depends on the number of digits in the fractional part. In this case, there were four, so the denominator is $10^4 = 10,000$. Similarly,

$$123.45 = 123 + \frac{45}{100} \quad \text{and} \quad 3.141592 = 3 + \frac{141,592}{1,000,000}$$

We shouldn't fear zeros in the fractional part as well, as long as they are not trailing. For example,

$$123.10043 = 123 + \frac{10043}{100,000} \quad \text{and} \quad 123.00043 = 123 + \frac{43}{100,000}$$

However, trailing zeros are not meaningful,

$$123.4900 = 123 + \frac{49}{100}$$

The trailing zeros in the fractional part don't add anything, just as leading zeros for the integer part are not meaningful.

In truth, we must be a bit careful here. You might very well see something like 123.4900 in a science book. There, the two trailing zeros might be meaningful. The number 123.4900 might be a measurement, in which case the trailing zeros indicate that we actually know that there are no thousandths or ten thousandths in the measurement. However, from the point of view of mathematics, trailing zeros in the fractional part are not meaningful. In a sense, there are an infinite number of trailing zeros for every decimal number. We'll return to this idea later in the chapter.

Let's recap. We've learned the following about decimal numbers,

1. Decimal numbers have three parts: the integer part, the decimal point, and the fractional part.

2. The fractional part of a decimal number is shorthand for a set of fractions, each one the digit value times a negative power of 10 beginning with -1 and becoming more negative as we add digits to the right.

3. Each position in the fractional part of a decimal number is 10 times larger than the position immediately to the right. This is also the case for the integer part.

4. The fractional part can always be written as a fraction by using a denominator of 10^n where n is the number of digits in the fractional part.

10.1.2 Speaking Decimals

How do you say 3.141592 out loud? Here's how I say it: "Three point one four one five nine two." We say the integer part as always, then the word "point" followed by a listing of the fractional digits in order.

If we need to refer to a specific digit, like the 5 in 3.141592, we say it's place value, in this case, the ten-thousandths place. Notice, we use the "ths" ending as we do with fractions. The names of the place values are off by one between the integer part and the fractional part. In 12,345.6789, the ten thousands place is occupied by the 1, while the ten-thousandths place by the 9. This is merely a consequence of how people have agreed to write decimal numbers. Frankly, it's always bothered me.

If instead of using a decimal point, we were to underline the 10^0 position like so, 123456789, then symmetry would be restored. The fourth place to the left would be the ten thousands, and the fourth place to the right the ten thousandths. Unfortunately, we must make do with what we have. Of course, you are free to write decimal numbers any way you wish in the privacy of your own home.

10.1.3 On the Number Line

Where do decimal numbers live on the number line? Everywhere. The decimals cover *all* of the number line, all the spaces between the fractions (rationals), and all the integers as well.

Mathematicians don't actually talk about "decimals" as that word refers to the notation used to write numbers. Instead, they speak of *real numbers*, and real numbers include all the types of numbers we've encountered: counting (natural) numbers, whole numbers, integers, rationals, and even the irrationals like $\sqrt{2}$, those numbers

10.1. CONCERNING DECIMALS

that cannot be written as fractions. Every possible number, every position on the number line, corresponds to a real number.

Let's look at a number line with some specific numbers marked on it. In most cases, the position is approximate. My goal is to help you imagine where different decimal numbers live and build your intuition about whether they are close to one integer or another. Here's the number line,

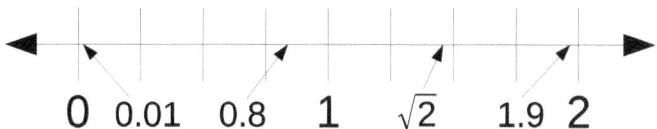

For example, look at 0.01. This number is one hundredth, 1/100. So, if we divide the distance from 0 to 1 into 100 equal-sized pieces, 0.01 covers one of them. Therefore, we expect 0.01 to be quite close to 0, as shown. Likewise, $0.8 = 8/10 = 4/5$. We expect, then, that 0.8 will be close to 1 because it will cover 4 out of 5 parts if we divide 0 to 1 into five pieces.

What about 1.9? This number is 9/10 of the way from 1 to 2, so we expect it to be close to 2.

I put $\sqrt{2}$ on the number line as well. As we'll learn later in the chapter, because $\sqrt{2}$ is irrational, we can only approximate its value with decimal numbers. As shown, the square root of 2 is approximately 1.414, meaning it's close to halfway between 1 and 2.

Thinking about where decimal numbers live on the number line is helpful, especially when we want to approximate the answer to a problem. For example, 1.990454×2 will take a bit of effort to calculate, but if we realize that 1.990454 is virtually 2, we get an approximate answer quickly: $2 \times 2 = 4$. The correct answer is 3.980908, which is very close to 4. Unfortunately, we don't have space in this book to spend much time discussing how to quickly approximate answers, but we'll revisit the subject briefly in Chapter 11 when we discuss how to calculate tips.

The real numbers, which mathematicians represent by the symbol $\mathbb{R}$, adds a final outer doll to our collection of Russian dolls[1],

---

[1]There is one more doll enclosing the reals, the complex numbers ($\mathbb{C}$) briefly mentioned in Chapter 9. Complex numbers have the form $a + bi$ for a and b real numbers and i the imaginary number. If $b = 0$, the reals are recovered.

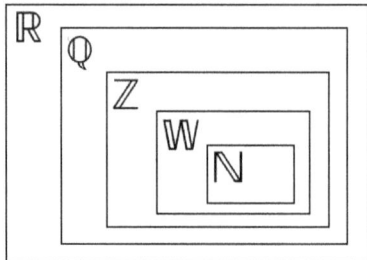

Read the diagram from the center out to put numbers into their relationships:

- The natural numbers (ℕ) are 1, 2, 3, etc.
- There is a whole number (𝕎) that is not a natural number, zero.
- There are integers (ℤ) that are not whole numbers, the negatives.
- There are rationals (ℚ) that are not integers. For example, 2/3.
- And, finally, there are reals (ℝ) like $\sqrt{2}$ that are not rational.

It's certainly not necessary to know or think about these relationships to learn the mechanics of working with decimal numbers. However, merely learning how to manipulate numbers shouldn't satisfy you. Mathematics is not magic, but it is magical, and to appreciate that fact, you need to know something of the background, something of what is behind the mere mechanics.

That said, mechanics are important, so let's now turn to arithmetic with decimal numbers.

10.1.4 Section Summary

This section introduced the concepts behind decimal numbers. First, we learned about the three parts that make up a decimal number: integer part, decimal point, and fractional part. We then realized that the fractional part continues place notation using negative powers of ten.

After a quick comment about how to speak a decimal number out loud, we examined where the decimal numbers live on the number line and learned that they *are* the number line, every position is a decimal number.

Finally, we learned that the decimal numbers are called real numbers by mathematicians and that they form a collection (set) of numbers that includes all the other number types we've explored in this book.

10.2 Adding and Subtracting Decimals

We add and subtract decimal numbers as we do integers with one additional requirement: *when writing the problem vertically, line up the decimal points*.

For example, lets add 23.45 and 9.04. First, write the problem vertically lining up the decimal points,

$$\begin{array}{r} 23.45 \\ +9.04 \\ \hline . \end{array}$$

Notice, I put the decimal point in the answer below the other decimal points.

That's it, we're done, at least as far as working with decimals is concerned. Now, we add as before ignoring the decimal point,

$$\begin{array}{r} 1 \\ 23.45 \\ +9.04 \\ \hline \mathbf{32.49} \end{array}$$

Notice the carry from adding 3 + 9. Here's another example, 83.7 + 12.88,

$$\begin{array}{r} 83.7 \\ +12.88 \\ \hline . \end{array}$$

Here, I line up the decimal points even though 83.7 has no value in the hundredths place. That's okay. Line up the decimal points, always. If one number doesn't have as many digits after the decimal point as the other, the missing trailing digits are zero. Therefore, the first addition we'll do is 0 + 8 = 8,

$$\begin{array}{r} 83.7 \\ +12.88 \\ \hline .8 \end{array}$$

Next, we add the tenths: $7 + 8 = 15$. We have a carry, is that a problem? No, because, by design, each digit position in the decimal rolls over as before. Ten tenths is $10/10 = 1$, so when the tenths place exceeds 9, we carry as before. This is true for all places after the decimal. Once we line up the decimal points, we add as before, including carries. Therefore, we write,

$$\begin{array}{r} 1 \\ 83.7 \\ +12.88 \\ \hline .58 \end{array}$$

The carry moves over to the ones column. To finish the problem, add the remaining digits,

$$\begin{array}{r} 1 \\ 83.7 \\ +12.88 \\ \hline \mathbf{96.58} \end{array}$$

Subtraction works in exactly the same way. Line up the decimal points, then subtract as always,

$$\begin{array}{r} 64.99 \\ -32.27 \\ \hline \mathbf{32.72} \end{array}$$

Likewise, borrowing works as before. For example, here's $41.73 - 9.65$,

$$\begin{array}{r} 41.73 \\ -9.65 \\ \hline . \end{array}$$

First, we need $3 - 5$ which produces a borrow from the next column to the left, the tenths,

$$\begin{array}{r} 6\,13 \\ 41.\not{7}\,\not{3} \\ -9.6\,5 \\ \hline .8 \end{array}$$

We can borrow from the tenths and give it to the hundredths because $1/10 = 10/100$. I.e., the tenths place is ten times as big as the hundredths place, just as we have with the integer part. Again, once we line up the decimal points, we subtract as always, ignoring the decimal points. Completing the problem gives us,

10.2. ADDING AND SUBTRACTING DECIMALS

$$\begin{array}{r} 3\;11 \\ 6\;13 \\ 4\,\rlap{/}1\,.\,\rlap{/}7\,3 \\ -\;\;\;9\,.\,6\,5 \\ \hline \mathbf{3\,2\,.\,0\,8} \end{array}$$

Please examine the problem above carefully to convince yourself that each step makes sense. Do notice the second borrow when subtracting the integer part.

What should we do with this problem?

$$\begin{array}{r} 12.0009 \\ -\;\;1.61 \\ \hline . \end{array}$$

We lined up the decimal points. The second number doesn't have enough digits, so mentally, we replace the spaces with zeros as if the problem were this,

$$\begin{array}{r} 12.0009 \\ -\;\;1.6100 \\ \hline . \end{array}$$

Now we subtract ignoring the decimal point. The first digit is $9-0 = 9$, easy enough. The next is $0-0 = 0$, also easy. So, at this point we have,

$$\begin{array}{r} 12.0009 \\ -\;\;1.6100 \\ \hline .\;\;\;09 \end{array}$$

Next, we need $0-1$, so we borrow from the tenths. However, the tenths is also 0, so we borrow from the ones to make the tenths 10, then immediately from the tenths to give to the hundredths,

$$\begin{array}{r} 1\;\;9\,10 \\ 1\rlap{/}2.\rlap{/}0\,\rlap{/}0\,09 \\ -\;\;1.6\,1\,00 \\ \hline .\;\;\;09 \end{array}$$

We're now in a position to complete the problem like so,

$$\begin{array}{r} 1\;\;9\,10 \\ 1\rlap{/}2.\rlap{/}0\,\rlap{/}0\,09 \\ -\;\;1.6\,1\,00 \\ \hline \mathbf{10.3\,9\,09} \end{array}$$

To recap, the general process to add or subtract two decimal numbers is,

- Write the problem vertically, lining up the decimal points and writing the decimal point below in the answer.
- Add or subtract as before, ignoring the decimal point.
- If one of the numbers has fewer digits in the fractional part, add trailing zeros as needed.

You may be wondering about negative decimals. Addition and subtraction of decimals follows the same rules for signs as adding and subtracting integers; nothing changes in that regard.

10.2.1 Section Summary

This section taught us that adding and subtracting decimal numbers is no more difficult than adding and subtracting integers provided we line up the decimal points.

10.3 Comparing Decimal Numbers

How do we know one decimal number is larger or smaller than another? For example, which number is bigger, 100.0934 or 100.1?

For integers, to know which number is bigger, we compare from left to right, considering each place value. If the current place value between the two numbers is the same, we move to the next place to the right. If those digits are the same, we move again to the right. We continue this process until we find two digits that don't match. Then, whichever one of those digits is larger tells us which number is larger.

For example, to compare 12345 and 12346, we start on the left with the 1, see that it matches, then move to the 2, see a match, move to the 3, etc. until we get to the ones column. Here we see that the first number has a 5 while the second has a 6. As $6 > 5$, we now know that the second number is larger than the first.

The same process happens with decimals. For the example above, we might line up the numbers vertically by aligning the decimal points, then scan left to right looking for a pair of digits that don't match,

$$100.0934$$
$$100.1$$
$$\uparrow$$

10.4. MULTIPLYING DECIMALS

The arrow marks the first place where the digits don't match. The top number has a 0, but the bottom has a 1. As $1 > 0$, we now know that $100.1 > 100.0934$.

What if the numbers are positive or negative? In those cases, if the signs are different, we immediately know the relationship. For example,

$$100.1 > -100.3809$$
$$102.275 > -1.4$$
$$0 > -0.0009$$
$$0.00001 > 0$$

Pay close attention to the last two examples. For $-0.0009 < 0$, we know this is true because *all* negative numbers, no matter how small, are less than zero. That's what makes them negative. Likewise, no matter how small, any positive number is greater than zero, which explains the final example.

10.4 Multiplying Decimals

Adding and subtracting decimals is easy enough, just line up the decimal points and proceed normally. Multiplying two decimals is, in a way, even more straightforward. Multiply as if the decimal points weren't there. After the multiplication is done, a bit of counting will tell you where the decimal point belongs in the answer.

Let's jump right in and multiply 16.183×4.1. We work vertically, but this time, we line up the end of each number, ignoring the decimal points altogether,

$$\begin{array}{r} 16183 \\ \times \quad 41 \\ \hline \end{array}$$

We know how to do this multiplication. The first partial product is 16183×1 and the second is 16183×4 but written beneath the 4. This gives us,

$$\begin{array}{r} 16183 \\ \times \quad 41 \\ \hline 16183 \\ +64732 \\ \hline 663503 \end{array}$$

Take a few moments to work the multiplication yourself on scrap paper to make sure I'm not lying.

So far, nothing new. Here's how we finish the problem. First, we need to put the decimal point in the answer. To find where, we *count* how many digits there are in the fractional parts of *both* numbers. The first number has three, 0.183, and the second one, 0.1, for a total of four. Therefore, we count four places in from the *right* of the answer and put the decimal point *before* that number, so we have four digits in the answer's fractional part.

For the problem above, we have 663503 after multiplying. Counting four in from the right puts us at the 3 in front of the 5. Therefore, the decimal point goes *before* the 3 to give us the final answer,

$$16.183 \times 4.1 = 66.3503$$

We can check by approximating. 16.183 is almost 16, and 4.1 is almost 4, so we expect the product to be close to $16 \times 4 = 64$, which it is.

Let's work another example. In this case, we'll multiply a decimal by an integer, 4.0922×3,

$$\begin{array}{r} 2 \\ 40922 \\ \times 3 \\ \hline 122766 \end{array}$$

To finish, we need to place the decimal point. There are four digits in the fractional part of 4.0922, and there are no digits in the fractional part of 3, so again, we count in four digits from the right to get to the second 2. Placing the decimal point before the digit gives us the final answer,

$$4.0922 \times 3 = 12.2766$$

To check, we note that 4.0922 is very close to 4 and that $4 \times 3 = 12$, so we expect an answer close to 12, like 12.2766.

A few more examples ought to do it. First consider this one where I'm multiplying 0.375 by 8,

$$\begin{array}{r} 375 \\ \times 8 \\ \hline 3000 \end{array}$$

10.4. MULTIPLYING DECIMALS

Notice, I ignore the zero in the integer part. We can ignore leading zeros in the integer part and trailing zeros in the fractional part.

There are three digits after the decimal point for 0.375 and none for 8, so we count in three digits from the right and put the decimal point before to get,

$$0.375 \times 8 = 3.000 = 3$$

I'm perfectly within my rights to change 3.000 to 3 as all zeros in the fractional part means there are no tenths, no hundredths, and no thousandths. If there aren't any, then no need to write them. Note, this problem multiplied a decimal by an integer and got an *integer* as a result. Hmm. We'll return to this example in the next section from a different perspective.

One more example. What about 0.001618×0.0037? Let's think carefully about this one. How many digits are meaningful in this case? I said that leading zeros for the integer portion do not count. Here, we have leading zeros for both numbers; the integer portion is zero. Also, for the fractional part, trailing zeros don't matter. Here we have none. So, for each number, the two zeros between the decimal point and the first nonzero digit, the 1 or the 3, *do* matter. We need to count them when we're done multiplying to know where to place the decimal point.

Our set up is,

$$\begin{array}{r} 1618 \\ \times 37 \\ \hline \end{array}$$

Multiplying as usual gives us,

$$\begin{array}{r} 12 \\ 415 \\ 1618 \\ \times 37 \\ \hline 11326 \\ +4854 \\ \hline 59866 \end{array}$$

All that remains is to put the decimal point in the proper place, so we count. From 0.001618 we get six decimals, and for 0.0037, we get four more, ten total. Therefore, we count ten from the right beginning with the rightmost digit of the answer. Note, we don't have ten digits in the answer, but that's okay. We use zeros for the digits we don't have. So, recalling that the decimal point goes before the digit, we count like so using zero for digits we don't have,

$$0.0000059866$$

Therefore,

$$0.001618 \times 0.0037 = 0.0000059866$$

Is the result a large number or a small number? It's a very small number. As a fraction,

$$\frac{59,866}{10,000,000,000} = \frac{29,933}{5,000,000,000}$$

This fraction is approximately 3 out of 500,000. Quite small, indeed.

Let's conclude this section by learning how straightforward it is to multiply any decimal number by a power of ten.

10.4.1 Multiplying by Powers of Ten

Multiplying an integer by a power of ten means adding zeros. For example,

$$123 \times 10^3 = 123 \times 1000 = 123,000$$

where multiplying by $10^3 = 1000$ added three zeros. In general, add an exponent's worth of zeros to the end of the integer.

Another way to think about the problem involves moving the decimal point. The decimal point for an integer is to the right of the ones column,

$$123 = 123.0$$

Here, I'm explicit in placing the decimal point and adding a zero in the tenths to make it easier to see the decimal point, much like we did above for numbers with only a fractional part like 0.007.

Where is the decimal point for 123,000? It's also to the right of the ones column, the rightmost zero. To go from 123.0 to 123000.0, we can imagine moving the decimal point three positions to the right and filling in the gaps with zero,

$$123000.0$$

However, this is precisely what multiplying by 1000 did. Therefore, *multiplying a decimal by a power of ten moves the decimal point the exponent's worth of positions to the right.*

10.5. CONVERTING FRACTIONS TO DECIMALS

For example, what's 123.007 × 1000? There are three zeros in 1000 because $10^3 = 1000$, so we move the decimal point three positions to the right,

$$123.007 \times 1000 = 123{,}007.0$$

Contemplate these examples, and I think you'll understand what I mean,

$$\begin{aligned} 34.056 \times 1000 &= 34{,}056.0 \\ 0.00031415 \times 10^4 &= 3.1415 \\ 0.00001 \times 100{,}000 &= 1.0 \\ 543.21 \times 10^7 &= 5{,}432{,}100{,}000.0 \end{aligned}$$

Notice, the last example moved the decimal point seven places to the right, which required adding zeroes to fill in the gaps, precisely what we did in going from 123.0 to 123,000.0.

We now know how to multiply decimals. I owe you an explanation of *why* the algorithm introduced in this section works. Before I can explain, we need first to discuss the division of decimals, and to do that, we need to know how to turn fractions into decimals.

10.4.2 Section Summary

Multiplication of two decimal numbers was the topic of this section. We learned the algorithm: ignore the decimal points, multiply as usual, then place the decimal point in the answer by counting from the right as many digits as are in both numbers' fractional parts. Along the way, we learned that multiplying by a power of ten is no more complicated than moving the decimal point to the right.

10.5 Converting Fractions to Decimals

We discussed fractions in Chapter 8. There, I said that among a fraction's many disguises is the one where it is a division problem. We didn't work the problem expressly then, but we'll do so now as that's how you turn a fraction into a decimal. Therefore let's turn 3/8 into a decimal by dividing 3 by 8,

$$8 \overline{\smash{)}3.0}^{\,\cdot}$$

I set up the problem as before but added a decimal point and a zero after it. As with addition and subtraction, when dividing a decimal, copy the decimal point from the dividend to the quotient, then forget about it.

Division now proceeds as before. We ask, "how many times does 8 go into 3?" and get zero as the answer. So, we look to the zero after the decimal point and ask, "how many times does 8 go into 30?" I'm ignoring the decimal point and treating the two digits together as 30.

The answer is 3 because $3 \times 8 = 24$ but $4 \times 8 = 32$. Therefore, we put 3 in the quotient above the 0 and subtract the 24,

```
       0.3
    8)3.0
      2 4
        6
```

Notice, I put the initial zero in the quotient. That's not necessary, but it helps in reading the value. Also, as we're using a fixed-spaced font for the long division process, I skipped a space to move past the decimal point in the dividend when subtracting $30 - 24 = 6$.

Are we done? No. When converting a decimal to a fraction, we're done when one of two things happens. The first is we get a remainder of zero. Six isn't zero, so we're not done yet. We'll get to the second possible outcome shortly.

Great, we're not done, but we're out of digits. Or are we? Decimal numbers have an infinite number of zeros after the last decimal digit. Each zero means that the corresponding negative power of 10 isn't present. We'll use this fact to pull down another zero, like so,

```
       0.3
    8)3.00
      2 4
       60
```

Now, we ask "how many times does 8 go into 60?" The answer is 7 because $7 \times 8 = 56$ and that's as close to 60 as we can get without going over. So, we add 7 to the quotient and subtract,

10.5. CONVERTING FRACTIONS TO DECIMALS

$$
\begin{array}{r}
0.37 \\
8\overline{)3.00} \\
\underline{24} \\
60 \\
\underline{56} \\
4
\end{array}
$$

We still don't have a remainder of zero. So, we pull another zero out of the ether and bring it down,

$$
\begin{array}{r}
0.37 \\
8\overline{)3.000} \\
\underline{24} \\
60 \\
\underline{56} \\
40
\end{array}
$$

How many 8's go into 40? Five as $5 \times 8 = 40$. So, we add a new digit to the quotient and subtract,

$$
\begin{array}{r}
0.375 \\
8\overline{)3.000} \\
\underline{24} \\
60 \\
\underline{56} \\
40 \\
\underline{40} \\
0
\end{array}
$$

At last, we have a remainder of zero, so we are done, and we now know that,

$$\frac{3}{8} = 0.375$$

Let's try another one, how about 1/3? Here's the set up,

$$
\begin{array}{r}
\cdot \\
3\overline{)1.0}
\end{array}
$$

Try this one on your own. If you get a remainder that isn't zero, just pull another zero down from the trailing part of the dividend as we did above. Stop whenever you are done or when you notice something fishy going on. No rush, I have all the time in the world.

Ready? Let's work it out together. We know that 3 doesn't go into 1, so we use the first zero after the decimal point and ask how many times 3 goes into 10. The answer is $3 \times 3 = 9$, so 3 goes in the quotient and we subtract 9,

$$
\begin{array}{r}
0.3 \\
3{\overline{\smash{\big)}\,1.0}} \\
\underline{9} \\
1
\end{array}
$$

The remainder isn't zero, so let's pull down another zero from the quotient and continue. We have 10, and we already know that $3 \times 3 = 9$, so we add another 3 to the quotient and subtract 9,

$$
\begin{array}{r}
0.33 \\
3{\overline{\smash{\big)}\,1.00}} \\
\underline{9} \\
10 \\
\underline{9} \\
1
\end{array}
$$

Hmm. We again have a remainder of 1. If we bring down another zero, we'll have 10 and again get 3 in the quotient and another remainder of 1. The process will *never* end. We'll never get a remainder of 0, only 1 after 1 after 1 forever.

This is the second possible outcome when turning a fraction into a decimal. If the decimal never ends, never comes up with a remainder of zero, then, eventually, the digits in the quotient will start to repeat. For 1/3, the digits repeat after the first one. The quotient is,

$$0.3333\ldots = 0.\overline{3}$$

The bar over the 3 is a new bit of notation. It means that the 3 repeats forever.

Let's recap. Every attempt to convert a fraction into a decimal ends in one of two ways:

1. The decimal will terminate when a remainder of zero is found.
2. The decimal will never terminate, and a portion of the quotient will repeat, over and over, forever.

10.5. CONVERTING FRACTIONS TO DECIMALS

Those are the only two options. Once part of the quotient starts to repeat, it will forever.

What's going on? In the first case, it means that the fraction can be represented by adding a set of fractions with powers of 10 as the denominators. For 3/8 we found,

$$\frac{3}{8} = 0.375 = \frac{3}{10} + \frac{7}{100} + \frac{5}{1000}$$

The sum of the three fractions on the right equals 3/8.

In the second case, it is impossible to find a set of fractions with denominators that are powers of 10 that add up to the initial fraction. For example, we learned that we could not find a set of fractions with denominators that are powers of 10 that add to 1/3. Recall, the fractional part of a decimal is precisely that, a compact way to represent a set of fractions with denominators that are powers of 10. That's why the *decimal expansion* of 1/3 repeats. When we convert a fraction to a decimal, we call the result the decimal expansion of the fraction. The decimal expansion will either terminate or repeat forever.

Another way to see that the decimal expansion will either terminate or repeat is to consider the process used in long division. Every time we subtract, we get a remainder, but not just any remainder, a remainder that must be less than the divisor. If that remainder is zero, we're done, and the expansion terminates. Otherwise, eventually, we'll hit a remainder that we already calculated. Once that happens, we're stuck in a cycle; the sequence of remainders, and their corresponding quotient digits, will repeat forever after that point.

Does this mean *all* decimal numbers terminate or repeat? No. Only those that are rationals, meaning the ratio of two integers, terminate or repeat. Irrational numbers, like $\sqrt{2}$ and $\sqrt{3}$, have decimal expansions that never terminate nor repeat.

Let's try a few more examples. What happens when we try to find the decimal expansion of 2/3? To save space, here's where we end up after two digits of the quotient have been found,

```
      0.66
   3)2.00
     1 8
       20
       18
        2
```

From this much of the expansion we can see that we'll always add a 6 to the quotient, therefore,

$$\frac{2}{3} = 0.\overline{6}$$

Here's a curious thing: 2/3 is twice 1/3, and adding $0.\overline{3}$ to itself is $0.\overline{6}$, i.e., we'll get a 6 in every digit position, but that's 2/3. Then what is 3/3? Shouldn't it be $0.\overline{3}$ added to itself three times?

$$0.\overline{3} + 0.\overline{3} + 0.\overline{3} = 0.\overline{9}$$

However, we *know* that 3/3 is an alias for 1. Therefore,

$$0.99999\ldots = 0.\overline{9} = 1$$

What?!? The statement above *is* correct. I'll use two arguments to help you believe that I'm not lying. The first appeals to your intuition, the second uses a tiny bit of algebra (gasp!), but I'm confident you'll follow along.

What's the difference between 1 and 0.9? It's 0.1. Fine. Now, what is the difference between 1 and 0.99? It's $1 - 0.99 = 0.01$. Let's keep adding digits to the second number and subtract it from one. We get,

$$1 - 0.999 = 0.001$$
$$1 - 0.9999 = 0.0001$$
$$1 - 0.99999 = 0.00001$$
$$1 - 0.999999 = 0.000001$$

As we add more 9's, the difference gets closer and closer to 0 meaning the second number is closer and closer to 1. It isn't too hard to imagine that if we do this forever, the difference will become 0 meaning $0.\overline{9}$ must equal 1.

Here's the second argument using a bit of algebra. I'll show the steps, then explain them,

10.5. CONVERTING FRACTIONS TO DECIMALS

$$x = 0.\overline{9}$$
$$10x = 9.\overline{9}$$
$$10x = 9 + 0.\overline{9}$$
$$10x = 9 + x$$
$$9x = 9$$
$$x = 1$$

I'm going to label $0.\overline{9}$ as x. That's the first line. The second line says that if I multiply $0.\overline{9}$ by 10, I'll move the decimal point one position to the right to get $9.\overline{9}$. As the 9's repeat forever, I still have the bar over the 9. On the other side of the equals sign, I must also multiply x by 10, that's $10x$.

The third line uses the fact that a decimal number is the sum of the integer part, here 9, and the fractional part, here $0.\overline{9}$. The fourth line replaces $0.\overline{9}$ with its label, x.

The fifth line subtracts the x from both sides of the equals sign. As multiplication is repeated addition, $10x$ means add x to itself 10 times. So if I take one x away, I'll have x added to itself nine times, which is $9x$. On the right side of the equals sign, the x goes away altogether. This leaves the equation $9x = 9$.

The equation $9x = 9$ is asking: what times 9 gives me 9? We know the answer; it's 1. Therefore, it must be that $x = 1$, and, as $x = 0.\overline{9}$, we have $0.\overline{9} = 1$.

It isn't the case that every repeating decimal expansion starts to repeat immediately, or that it has only one repeating digit. For example, and I leave working them out as exercises for you, consider these repeating fractions,

$$\frac{1}{6} = 0.1\overline{6}$$

$$\frac{1}{7} = 0.\overline{142857}$$

$$\frac{1}{19} = 0.\overline{052631578947368421}$$

I don't recommend working out 1/19 as a repeating decimal by hand. I mean, you can, if you want to, but, seriously, I wouldn't bother.

1/2	0.5	1/9	0.$\overline{1}$
1/3	0.$\overline{3}$	1/10	0.1
1/4	0.25	1/12	0.08$\overline{3}$
1/5	0.2	1/16	0.0625
1/8	0.125	1/20	0.05

Table 10.1: Common unit fractions and their decimal equivalents.

Your calculator won't help here, either, as virtually all common calculators lack enough precision. However, the standard Unix desk calculator program, dc, will calculate with enough precision. The sequence "54 k 1 19 / p" tells dc to evaluate 1/19 using 54 digits after the decimal point producing,

.052631578947368421 052631578947368421 052631578947368421

thereby proving my claim correct.

Table 10.1 lists common fractions and their decimal equivalents. You'll run across these fractions frequently, so it's worth remembering their decimal representation. You might even work some of them out yourself on a piece of scrap paper. Most non-metric rulers use 1/2, 1/4, 1/8, and often 1/16 to divide the spaces between inches.

10.5.1 Section Summary

This section taught us how to convert fractions into decimal numbers by using long division. We learned that we could bring down zeros repeatedly to continue adding digits to the quotient. We also learned that eventually, the process terminates with a remainder of zero or continues forever, repeating a portion of the quotient.

10.6 Dividing Decimals

In this section, we work our way up to dividing decimals by other decimals. First, we learn that dividing by powers of 10 is just as easy as multiplying by powers of 10. Second, we use our new superpower of dividing by powers of 10 to understand why the multiplication algorithm we learned above actually works. Third, we learn that dividing decimals by integers is as easy as converting a fraction to a decimal. Finally, we conquer dividing decimals by decimals. Let's get to work.

10.6.1 Dividing by Powers of Ten

To multiply by a power of ten, we move the decimal point to the right the exponent's number of places. For example, multiplying by $1000 = 10^3$ moves the decimal point three places to the right, filling gaps with zeros.

As you might guess, to divide by a power of 10, we move the decimal point the other way; we move it *to the left*. In truth, this is nothing new. We've been doing this all through the chapter, every time we multiplied two decimal numbers together. The part of the algorithm where we count from the right to decide where the decimal point goes is just that, dividing by a power of 10. We'll see this explicitly below.

Above, we multiplied 123 by 1000. Let's now divide instead. Graphically, this means we move the decimal point three positions to the left,

$$0.123$$

A few additional examples to demonstrate the process,

$$54321 \div 10^4 = 5.4321$$
$$34.056 \div 1000 = 0.034056$$
$$0.1234 \div 100 = 0.001234$$
$$543.21 \div 10^6 = 0.00054321$$
$$543.21 \div 10^{-6} = 543{,}210{,}000$$

Let's think carefully about the final example above. We are dividing by a negative power of 10. Recall, Chapter 9 presented us with similar examples where we added or subtracted exponents because we were working with only powers of 10.

The case above is division, but division means subtracting exponents, but the exponent is negative. Just as subtracting a negative implies addition, so division by a negative power of ten implies moving the decimal point in the opposite direction, meaning to the right instead of the left. Likewise, multiplying by a negative power of 10 means to move the decimal point to the left,

$$543.21 \times 10^{-6} = 0.00054321$$

Notice, we find the same answer as $543.21 \div 10^6$, which shouldn't really surprise us since we know that

$$10^{-6} = \frac{1}{10^6}$$

meaning multiplying by 10^{-6} is the same as dividing by 10^6. Read exponents carefully.

To summarize, division of a decimal by a power of 10, e.g., 10^n for some integer n, means,

- Move the decimal point n places *to the left*.
- Add zeros to missing places as needed.
- If n is negative, move the decimal to the right instead.

Now that we know how to multiply and divide by powers of 10, we're finally in a position to understand why the decimal multiplication algorithm works the way it does.

Somewhere along the way, you may have encountered numbers that look like this,

$$6.022 \times 10^{23}$$

Here, a decimal, usually with only one digit in the integer part, is multiplied by ten raised to some positive or negative power. Numbers like this are examples of *scientific notation*. Scientists and engineers use scientific notation to keep from wasting time counting zeros and to make multiplication and division of numbers easier. This was especially true back in the days of slide rules.

Scientific notation writes very small numbers and very large numbers in the same way. For example, a very small number,

$$1.414 \times 10^{-18} = 0.000000000000000001414$$

and a very large number,

$$1.414 \times 10^{18} = 1,414,000,000,000,000,000$$

Computers use a slightly different form of scientific notation. For example, the numbers above are entered as `1.414e-18` and `1.414e18`, respectively.

Multiplication and division with scientific notation first multiplies or divides the decimal part, then adds or subtracts the exponents. Add the exponents if multiplying; subtract them if dividing. For example,

$$(2.5 \times 10^6)(4 \times 10^{-3}) = (2.5)(4) \times 10^{6+^-3} = 10.0 \times 10^3$$

and

10.6. DIVIDING DECIMALS

$$\frac{6 \times 10^8}{2 \times 10^3} = \left(\frac{6}{2}\right) \times 10^{8-3} = 3 \times 10^5$$

We won't work with scientific notation in this book, but I wanted to explain it a bit because you do see it from time to time, often when reviewing science results presented in the popular press.

10.6.2 Why The Multiplication Algorithm Works

Multiplying two decimal numbers means ignoring the decimal points. When done, we count in from the rightmost digit of the answer as many places as there were digits in the fractional parts of the numbers multiplied. We're now in a position to understand why this algorithm works.

Our first decimal multiplication example was:

$$16.183 \times 4.1 = 66.3503$$

I'm going to rewrite the problem using integers and powers of 10. Here it is,

$$16183 \times 10^{-3} \times 41 \times 10^{-1}$$

I'm multiplying four numbers. As we know, multiplication does not care about the order in which we multiply, $4 \times 2 = 2 \times 4 = 8$. This fact holds for any number of things multiplied together. Therefore, I can move numbers around like this,

$$16183 \times 41 \times 10^{-3} \times 10^{-1}$$

And, because I'm just that sort of person, I'm going to multiply the two powers of 10 by adding their exponents together, $-3 + {}^-1 = -4$. This gives us,

$$16183 \times 41 \times 10^{-4}$$

So they don't feel left out, I'll multiply the two integers as well giving,

$$663503 \times 10^{-4}$$

To finish, I need to multiply an integer by a power of 10. In this case, the exponent is negative, so I need to move the decimal point *four places to the left*. Doing so gives me 66.3503, which is precisely what we get when using the decimal multiplication algorithm.

The decimal multiplication algorithm works because it first finds the product of two integers, those that are made by moving the respective decimal points to the right. For 16.183, we get 16183 by moving three positions; for 4.1, we get 41 by moving one position. In essence, we've multiplied the product of 16.183 and 4.1 by $10^{3+1} = 10^4$. Therefore, to get the actual product, we need to divide the product of 16183 × 41 by 10^4. Above, we did this by the equivalent operation of multiplying by 10^{-4}.

There is nothing special about 16.183 and 4.1, at least that I'm aware of, so this moving of decimal points to the right followed by moving them again to the left when done is valid for any two decimal numbers. This is why the decimal multiplication algorithm works as it does.

10.6.3 Dividing by Integers

We learned above how to convert fractions into decimals by dividing the numerator by the denominator. Now, let's generalize one level and explore dividing a decimal number by an integer. We begin with an example, 3.5556 ÷ 12. We set up the division problem as you might expect,

$$12\overline{)3.5556}$$

Notice, as when converting a fraction to a decimal, we write the decimal point in the quotient directly above the decimal point in the dividend. And, that's it. We ignore the decimal point going forward and use long division as before. This example, worked completely, gives us,

```
        0.2963
   12)3.5556
       2 4
       1 15
       1 08
          75
          72
           36
           36
            0
```

10.6. DIVIDING DECIMALS

I recommend stopping here and copying the problem to a piece of paper so you can work it yourself; don't just take my word for it that $3.5556 \div 12 = 0.2963$ (even though it does).

What does $3.5556 \div 12 = 0.2963$ even mean? We're dividing by an integer, 12, so one way to think about this problem is that the distance from 0 to 3.5556 on the number line has been split into twelve equal-sized parts. How long is each part? It's 0.2963. And, as we now understand, writing "0.2963" means the fraction 2963/10,000, which is rather close to the fraction $3000/10,000 = 3/10$.

Let's work one more example together. How about $6.283 \div 2$? Give it a go first, then come back and see if you agree with my approach. Here's what I get,

$$
\begin{array}{r}
3.1415 \\
2{\overline{\smash{)}6.283}} \\
\underline{6} \\
02 \\
\underline{2} \\
08 \\
\underline{8} \\
03 \\
\underline{2} \\
10 \\
\underline{10} \\
0
\end{array}
$$

This example did do something new. After the final subtraction, $3 - 2 = 1$, I pulled down a zero and continued the process. We did much the same when converting fractions to decimals. As in that case, we can do the same when dividing decimals by integers. We pull down trailing zeros as needed until we arrive at a remainder of zero.

The two examples in this section illustrate the general process, which is in the end, no different than the process used to turn a fraction into a decimal: set up the problem, copy the decimal point into the quotient, divide as before ignoring the decimal points, and bring down zeros as needed.

When converting a fraction to a decimal, we learned that we always end with a remainder of zero or find ourselves in a situation with a repeating sequence of decimal digits. Converting a fraction to an integer is dividing one integer by another. Will the same situation happen when dividing a decimal by an integer? Will we always terminate or find ourselves trapped in a repeating hell of decimal digits? Yes, because for any dividend that is a decimal number, say with n digits after the decimal, we can move the decimal n digits to the right to make it an integer. Doing this will only move the decimal in the quotient to the right. At that point, we have an integer divided by an integer, so we will terminate or end up repeating.

Another way to think about this is that all decimal numbers you can write down, meaning a finite number of digits after the decimal point, can be converted into a fraction; we saw this above. So, yes, when dividing by an integer, you will still find either a terminating decimal as the quotient or one that repeats some sequence of quotient digits forever.

10.6.4 Dividing by Decimals

How should we approach a problem like $23.974 \div 1.48$? Let me rewrite the problem using slash notation,

$$\frac{23.974}{1.48} = \left(\frac{23.974}{1.48}\right)\left(\frac{100}{100}\right)$$

On the left is the problem. It looks like a fraction, but it isn't; the numbers are not integers. However, it is a valid way to write $23.974 \div 1.48$. On the right, I've taken what is on the left and multiplied it by the 100/100 alias for 1. We know that multiplying a number by 1 does not change its value, so the left and right sides above are equal; they represent the same decimal number.

Fine, but what does multiplying by 100/100 buy us? Let's work out the multiplication, the top is multiplied by 100, as is the bottom,

$$\left(\frac{23.974}{1.48}\right)\left(\frac{100}{100}\right) = \frac{2397.4}{148}$$

To multiply a decimal by 100, we move the decimal point two places to the right. Now, look at the expression on the right of the equals sign. Whatever value we get by dividing 2397.4 by 148 is the same value we'd get by dividing 23.974 by 1.48. However, in the former case, we are now dividing by an integer, and we know how to do that.

That's the trick to dividing a decimal by another decimal: *move the divisor's decimal point as far to the right as possible. Then, move the dividend's decimal point that many places as well.* Doing this turns the problem into one of dividing a decimal number by an integer.

10.6. DIVIDING DECIMALS

Here are a few more examples. The important point here is the set up, the process of multiplying by the right alias for 1 so that the divisor is now an integer. Consider,

$$\frac{3.141592}{2.001} = \left(\frac{3.141592}{2.001}\right)\left(\frac{1000}{1000}\right) = \frac{3141.592}{2001}$$

$$\frac{65.02}{0.0054} = \left(\frac{65.02}{0.0054}\right)\left(\frac{10000}{10000}\right) = \frac{650200}{54}$$

$$\frac{4}{2.718} = \left(\frac{4}{2.718}\right)\left(\frac{1000}{1000}\right) = \frac{4000}{2718}$$

$$\frac{-0.7989}{256.64} = \left(\frac{-0.7989}{265.64}\right)\left(\frac{100}{100}\right) = \frac{-79.89}{25664}$$

The final example reminds us of negative numbers. As we might have come to expect by now, the proper sign for multiplying and dividing negative decimals follows the same rules as integers: if the signs are the same, the answer is positive; otherwise, the answer is negative.

What does $23.974 \div 1.48$ mean? In this form, it's hard to assign a meaning. However, once we scale it and realize that it refers to the same decimal number as $2397.4 \div 148$, we can visualize the meaning via the number line: take the distance from 0 to 2397.4, divide it into 148 equal-sized pieces, and keep one of those pieces.

On a practical note, if you try to work out most of the examples above, you'll quickly become bored or frustrated. Even though each example involves a decimal with a fixed number of digits, meaning the quotient will eventually terminate or repeat, the number of digits you need to calculate is far more than you'll care to spend the time calculating. The only exception is $65.02 \div 0.0054 = 12040.\overline{740}$, which repeats after three digits. For example, $4 \div 2.718$ becomes,

$$1.\overline{471670345842531272994849153789551140544518027961736571008094186902133922001}$$

and $3.141592 \div 2.001$ is even worse,

$$1.\overline{5700109945027486256871564217891054472763618190904547726136931534232883558220889555222388805597201399300349825087456271864067966016991504247876061969015492253873063468265867066466766616691654172913543228385807096451774112943528235882058970514742628685657171414292853573213393303348325837081459270364817591204397 8}$$

No one ever said that the repeating portion would be "nice." In practice, for all but the simplest of decimal divided by decimal problems, you're better off using a calculator.

10.6.5 Section Summary

The division of decimal numbers was the focus of this section. We first learned that dividing a decimal number by a power of ten means to move the decimal point to the left based on the exponent. With that knowledge, we were then able to learn why the decimal multiplication algorithm works as it does.

We followed by dividing decimal numbers by integers. The process is virtually identical to converting a fraction into a decimal: copy the decimal point into the quotient, then divide as normal bringing down zeros as necessary until the remainder becomes zero or a repeating sequence of quotient digits appears. Additionally, we learned that even though dividing a decimal by an integer will always terminate or repeat, the repeating portion need not be simple, nor short.

We concluded the section by learning that division of a decimal number by another decimal number gives the same answer as moving the decimal point in both the divisor and dividend as many places to the right as necessary to make the divisor an integer, then dividing.

10.7 Decimal Numbers and Computers

In general, when a computer stores or uses an integer, like 8675309, it is working precisely with that integer. This is because the computer can store that number exactly. However, decimal numbers can be unending in their decimal expansions. So how does a computer with limited memory store a number with an endless number of decimal digits? The answer is it can't. Instead, the computer approximates the number because it only uses a fixed number of bits (binary digits).

Earlier in the chapter, we solved 16.183 × 4.1 and found that it is 66.3503, exactly. However, when I ask a computer to do the multiplication, here's what I get,

$$16.183 \times 4.1 = 66.35029999999999$$

10.7. DECIMAL NUMBERS AND COMPUTERS

The computer cannot store the decimal numbers exactly; it must approximate them, leading to small errors cropping up in the calculations.

Most small errors are a wash. They average out or don't adversely influence the final result of the calculation. However, sometimes, this isn't the case. Sometimes, these *rounding errors* add up and lead to catastrophe.

During the Gulf War (August 1990 through February 1991), the Iraqi army deployed Soviet-built Scud missiles. In response, the United States installed Patriot anti-missile batteries to shoot down the Scuds.

On February 25, 1991, a Scud missile hit the US Army barracks in Dharan, Saudi Arabia, killing 28 soldiers. The base had a Patriot missile battery, but the battery failed to fire. An analysis of the battery's software revealed that a timer was incremented by adding 0.1 every tenth of a second. However, 0.1, when represented in binary, is a repeating value, not a terminating decimal as it is in base 10. Therefore, each increment of the timer introduced an error of about 0.000000095 seconds.

This is a tiny value, but for the 100 or so hours the battery was in constant operation, the error accumulated. When the Scud missile appeared, the timer was off by about 0.34 seconds. A Scud missile covers about 0.3 of a mile in about 0.34 seconds. The error was such that the battery believed the missile was out of range, so it did not fire, resulting in the soldiers' deaths.

Computers must work with limited memory and therefore limited precision, really no different from the calculations we've been exploring in this chapter. Ultimately, even the most advanced calculations performed by computers are built from the same four basic operations: addition, subtraction, multiplication, and division. Errors in calculations are inevitable when decimal numbers are used. In general, we are safe using software based on decimal (called *floating point*) calculations because standard software testing catches the majority of such errors. However, not always. Software engineers must evaluate their floating-point code carefully and thoroughly. Doubly so for critical software where lives are on the line.

Interestingly, modern deep neural networks, the kind powering the AI revolution, are different. Deep neural networks are robust to floating-point errors. In many cases, it is possible to reduce the precision of the network's parameters to only a handful of digits while not affecting the network's performance. That's good news for those concerned that a round-off error might transform a harmless power grid watchdog AI into Skynet.

10.8 Chapter Summary

This chapter explored decimal numbers, meaning numbers that have both an integer part and a fractional part expressed as a series of digits representing negative powers of ten.

The chapter opened by considering decimals numbers as a concept. We learned the parts of a decimal number, including the fractional part, by continuing place notation to include negative powers of ten. Next, we learned how to speak decimal numbers aloud and where they live on the number line.

The primary goal of the chapter is to learn how to do arithmetic with decimal numbers. We began with addition and subtraction, which work the same way as the addition and subtraction of integers once the decimal points are aligned. Then, after a brief aside about comparing the magnitude of decimal numbers, we learned the multiplication algorithm. Along the way, we learned that multiplying by powers of ten is particularly simple, just move the decimal point right for positive powers and left for negative powers.

Chapter 8 taught us about fractions. In this chapter, we learned how to express fractions as decimal numbers and discovered that the decimal expansion of fractions would either terminate or repeat indefinitely.

Division of decimals was tackled next. We started with division by powers of ten, which is as simple as moving the decimal point to the left for positive powers and to the right for negative powers, precisely the opposite of multiplication by powers of ten. With division by powers of ten under our belt, we then explored the "why" behind the decimal multiplication algorithm.

Division of decimals by an integer concerned us next. We learned that it is not different from integer long division once we copy the decimal point into the quotient and that we can "bring down" zeros as needed to continue the division until we get a terminating answer or a repeating set of quotient digits. To complete our exploration, we learned that the trick to dividing a decimal number by another decimal number is to multiply both by the alias for 1 that turns the divisor into an integer, then divide as usual.

We concluded the chapter by briefly considering how computers deal with decimal numbers and learned that, unfortunately, paying too little attention to the errors produced by restricting repeating decimals to a fixed number of digits in the fractional part can lead to severe consequences.

10.9 Terms and Concepts

We introduced the following terms and concepts in this chapter.

decimal expansion Writing a number in decimal form, in particular, converting a fraction into a decimal.

decimal number A number with an integer part, a decimal point, and a fractional part representing a series of fractions with denominators that are powers of ten.

decimal point A period used primarily in North America to separate a decimal number's integer and fractional parts.

decimal separator The more generic name for the decimal point. Sometimes referred to as the radix point, especially if a base other than ten is used.

floating point Another name for decimal numbers, but in particular use in computer science to refer to how computers attempt to represent decimal numbers.

real numbers The mathematician's name for decimal numbers. The set of real numbers includes all integers, rationals, and irrationals.

rounding errors Tiny errors computers are forced to commit because it is impossible to represent decimal numbers with infinite accuracy in a machine with a fixed-size memory.

scientific notation A number form made up of a decimal number, called the significand or mantissa, multiplied by 10 raised to some power. Scientific notation is used in science and engineering to simplify writing numbers and minimize possible errors during calculations.

10.10 Exercises

Exercise 1

Convert the following fractions into decimals. Work far enough to see the repeating portion of the expansion, if there is one.

a) $\dfrac{1}{11}$ b) $\dfrac{5}{8}$

c) $\dfrac{1}{13}$ d) $\dfrac{5}{6}$

e) $\dfrac{3}{7}$ f) $\dfrac{7}{15}$

Exercise 2

Replace □ with the proper symbol, either > or <.

a) 97.005 □ 97.004
b) 0.1011 □ 0.01
c) 101.01 □ 101.1
d) −0.123 □ 0.123
e) 0 □ −4.11
f) 0 □ 4.11

Exercise 3

Solve the following decimal multiplication problems.

$$\begin{array}{r} -0.2502 \\ \times 0.067 \\ \hline \end{array} \qquad \begin{array}{r} 0.5926 \\ \times -0.0015 \\ \hline \end{array} \qquad \begin{array}{r} 0.564 \\ \times -4.2 \\ \hline \end{array} \qquad \begin{array}{r} 0.1091 \\ \times 6.1 \\ \hline \end{array}$$

$$\begin{array}{r} -0.5397 \\ \times -0.028 \\ \hline \end{array} \qquad \begin{array}{r} 679.8 \\ \times 4.2 \\ \hline \end{array} \qquad \begin{array}{r} 0.0304 \\ \times -0.0064 \\ \hline \end{array} \qquad \begin{array}{r} -686.1 \\ \times 0.0095 \\ \hline \end{array}$$

$$\begin{array}{r} 3.13 \\ \times -0.44 \\ \hline \end{array} \qquad \begin{array}{r} 392.9 \\ \times 0.0011 \\ \hline \end{array} \qquad \begin{array}{r} -40.19 \\ \times 0.0052 \\ \hline \end{array} \qquad \begin{array}{r} -0.1201 \\ \times -0.92 \\ \hline \end{array}$$

$$\begin{array}{r} -241.1 \\ \times -0.083 \\ \hline \end{array} \qquad \begin{array}{r} 430.7 \\ \times -0.093 \\ \hline \end{array} \qquad \begin{array}{r} -5.832 \\ \times -0.063 \\ \hline \end{array} \qquad \begin{array}{r} 5.718 \\ \times 0.99 \\ \hline \end{array}$$

Think About It

You may have noticed while reading this chapter that a number like 3.141592 appeared several times. Perhaps you are already familiar with this number. It has a special name, π (pi), and it is the ratio between the circumference of a circle and its diameter. The circumference is the length around the circle. The diameter is the length of the line segment that goes from one side of the circle to the other, passing through the center. Specifically, $\pi = C/d$ for circumference C and diameter, d.

Pi is a good example of a decimal number. But, in reality, it is far more than that. Pi shows up everywhere in mathematics, science, and engineering. Even the ancients were aware of it, though they lacked the means of approximating it very well. Notice, I said "approximate it." Pi is an irrational number. Its decimal approximation never ends, nor does it repeat. For example, here's π to 44 digits,

$$3.14159265358979323846264338327950288419716939$$

Books could be, and many have been, written about π. People have fun memorizing as many digits of π as possible (not me). As of this writing, the current world record for correctly reciting digits of π from memory is over 100,000 digits.

Here are some fractions commonly used to approximate π,

$$\frac{22}{7} = 3.142857\ldots \qquad \frac{355}{113} = 3.141592\ldots$$

One amazing thing about the decimal expansion of π is that it's computable, meaning it's possible to keep calculating digits. Currently, π has been calculated to at least 62.8 trillion digits (62,800,000,000,000)!

Pi is believed to be what mathematicians call a *normal number*. A normal number is one where any desired set of n digits appears equally often in the decimal expansion. If this is so, then *any* set of digits, no matter how long, occurs somewhere in the decimal expansion of π, not just once, but *infinitely* many times. Let that sink in for a bit.

The website `https://mypiday.com/` will locate the first occurrence of your birthday in the expansion of π. The first eight digits of π are 14159265. This sequence happens again, beginning with the 191,525,092-th digit; π contains π.

Computers use a number code to represent characters. The traditional code is known as ASCII. For example, in ASCII, the computer represents the string "CAT" with the numbers 67, 65, and 84, or 676584. This sequence of numbers first occurs $2,760,758$ digits into the expansion of π. Likewise, "MATH" appears at digit $57,460,688$.

So, π can be interpreted as text. And, if normal, π contains an infinite number of copies of any sequence of characters at all. This means the entire text of every book in the world is already present in the expansion of π (or is it, perhaps not until someone finds it?) This includes your entire life story, no matter how your future biographer chooses to write it. Amazing, humbling, and just plain weird. Ain't math grand?

Use the website above to locate your birthday in π. Also, don't forget to celebrate Pi Day, March 14th, with a piece of your favorite pie.

Chapter 11

Percents

In this final chapter, we explore the world of *percents*, numbers representing fractions with a denominator of 100. Percents are extremely common, so understanding how to work with them is essential. Everything we've learned up to this point applies to percents as percents are represented both as fractions and decimals.

This chapter is more applied than previous chapters. We'll work practical problems, including those that require us to extract relevant information from descriptions, i.e., so-called *word problems*.

Specifically, we first explore the meaning of the word "percent." Doing so helps us to build an understanding of what a percent represents. We then dive into the relationship between fractions, decimals, and percents. This is necessary as percents are fractions usually expressed as decimals, so we must understand how to convert one to the other.

Next comes a series of sections on percents in action. Some are commonplace, like calculating a given percent of a number, the price of an item on sale, what to tip, or what the sales tax will be. Others relate more to the world of data, like calculating poverty rates or comparing two AI's to decide which is better. Still, others are concerned with compound interest, percent change, and the distinction between a percent and a percentage point.

By the end of the chapter, you'll know what a percent is, how to interpret it, and how to use it in practical situations. Along the way, I think you'll have some fun, too.

11.1 What Is A Percent?

Speakers of Romance languages, like Spanish, French, Italian, or Portuguese, likely already know the meaning of the word "percent." It comes from Latin. In Spanish, the words "por ciento" are translated into English as "percent." Individually, "por" means "for" or "by", and "ciento" means "hundred." So, we might interpret "percent" as meaning "for each hundred" or "by the hundred." Keep this in the back of your mind as you work through the chapter.

Percents are not a new kind of number, like decimals or rationals. Instead, percents are perhaps most easily thought of as fractions where the denominator is always 100. Typically, fractions divide the number line from 0 to 1 into denominator-sized chucks, then count the numerator's worth of them. I.e., 4/5 divides 0 to 1 into five parts and keeps four of them. A percent divides 0 to 1 into 100 parts instead.

To me, the easiest way to get a handle on percents is to first learn about their relationship to fractions and decimals. Once that is clarified, I suspect you'll get the idea. So, let's dive right in.

11.2 Fractions, Decimals, and Percents

Table 10.1 lists common unit fractions, those with a numerator of 1, and their decimal equivalents. For example, 1/4 = 0.25, according to the table. However, the previous chapter taught us how to turn a decimal into a fraction, so, putting all of this together, I can write a series of expressions that are all equal to each other,

$$\frac{1}{4} = 0.25 = \frac{25}{100} = 25\%$$

Notice, I added something new on the right side, 25%. This number is a percent. You know it's a percent because the number has a %, the percent sign, after it. The percent sign is a shorthand way of referring to the fraction 25/100. When you see %, think of the fraction. So,

$$\frac{57}{100} = 57\% \text{ and } \frac{23.7}{100} = 23.7\%$$

The example on the right tells us that percents are not necessarily rational numbers, the numerator need not be an integer; it's perfectly fine if it isn't. So, percents can be thought of as fractions with a denominator of 100, hence "by the hundred." A percent is a part of

11.2. FRACTIONS, DECIMALS, AND PERCENTS

a whole, but instead of the whole being viewed as 1, we're viewing it as 100. Therefore, all of something is 100% of it.

11.2.1 Fractions to Percents

How do we change a fraction into a percent? Let's turn 3/8 into a percent. That means we must find a number, call it x, that makes this equation true,

$$\frac{3}{8} = \frac{x}{100}$$

This equation is asking: *what number can we put in for x so that when it is divided by 100 we get the same number as dividing 3 by 8?*

To answer the question, we need to know what number we get when we divide 3 by 8. In the previous chapter, we learned how to turn a fraction into a decimal number: divide the numerator by the denominator. We did just this for 3/8 in Chapter 10, so we already know 3/8 = 0.375. However, if we don't want to do the long division, but remember from Table 10.1 that 1/8 = 0.125, then we get 3/8 by multiplying 1/8 by 3 giving us 3 × 0.125 = 0.375. This is why it's helpful to remember the decimal equivalents for common unit fractions.

Okay, we have 3/8 as a decimal. Now what? We've turned our problem into this,

$$\frac{3}{8} = 0.375 = \frac{x}{100}$$

We need to find a number, x, that, when divided by 100, gives us 0.375. Here's why percents use 100, because it is a power of 10, i.e., $100 = 10^2$. We know how to divide and multiply by powers of 10. We need this number: $x = 0.375 \times 100$. Multiplying by 100 means moving the decimal point two places to the right, so we now have our answer, $x = 37.5$ and we can write,

$$\frac{3}{8} = 0.375 = \frac{37.5}{100} = 37.5\%$$

All the words above boil down to two simple steps to convert a fraction into a percent:

1. Convert the fraction into a decimal by dividing the numerator by the denominator.

2. Multiply the decimal by 100, i.e., move the decimal point two places to the *right*.

That's it, that's how you convert a fraction into a percent. Let's try another example, 2/3. We know already that $2/3 = 0.\overline{6}$, we found this in the previous chapter, so we need only multiply $0.\overline{6}$ by 100,

$$\frac{2}{3} = 0.\overline{6} = 66.\overline{6}\%$$

Consider the expressions above for a bit until you are comfortable with the notion that even though $0.\overline{6}$ repeats forever; it's not only okay to multiply it by 100, but that you do indeed get $66.\overline{6}$. Recall, $0.6\overline{66}$ is a perfectly valid way to write $0.\overline{6}$.

To emphasize that percents are fractions, here's where certain percents live on the number line,

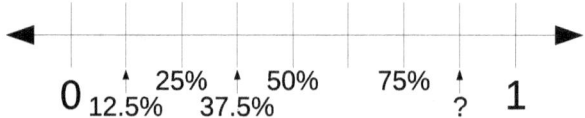

This is so because,

$$\frac{1}{8} = 12.5\%, \quad \frac{1}{4} = 25\%, \quad \frac{3}{8} = 37.5\%, \quad \frac{1}{2} = 50\%, \quad \frac{3}{4} = 75\%$$

What percent label should we give to the question mark? The number line is divided into eighths, each mark covers one-eighth of the distance from 0 to 1. The question mark is labeling the seventh mark, so it represents the fraction 7/8. To get 7/8 as a decimal, we multiply $1/8 = 0.125$ by seven,

$$7 \times 0.125 = 0.875 = 87.5\%$$

Therefore, the question mark is 87.5% of the way from 0 to 1. Also, the first interval, 1/8 is 12.5%. The last labeled mark is 75% and 75% + 12.5% = 87.5%, meaning percents add as you might expect them to.

Notice, percents, since they do not need to be integers, are a way to label any point between 0 and 1 with a decimal number from 0 to 100.

11.2.2 Decimals to Percents

To convert a decimal to a percent, multiply the decimal by 100. That's all. So, the following are all ways of writing the same number, first as a decimal, then as a percent,

$$0.125 = 12.5\%, \quad 0.14159 = 14.159\%, \quad 1.03 = 103\%, \quad 16.337 = 1633.7\%$$

Notice, percents are not restricted to a maximum of 100%. Percents greater than 100% are equivalent to improper fractions. Consider, 123% = 123/100, an improper fraction because the numerator is greater than the denominator.

Since a percent merely multiplies a decimal by 100, percents map all the decimals from [0, 1] to [0, 100]. If that's the case, why not use the fractions or their decimal equivalents? Why this extra multiplication by 100?

One answer is because humans generally don't like talking about small numbers in the range [0, 1]. Which is easier to understand: "fewer than 0.1 of people are left-handed" or "fewer than 10% of people are left-handed?" The second form is more accessible for most because we are used to numbers like 10, but not 0.1, even though the first form is also correct.

Additionally, as percents are by the hundred, i.e., 0.1 = 10/100, we can easily express 10% as "fewer than 10 out of 100 people are left-handed." Using a percent makes it easy to visualize things with charts as well. Say we knew that exactly 10% of people were left-handed. Then, we might visualize this fact with something like this,

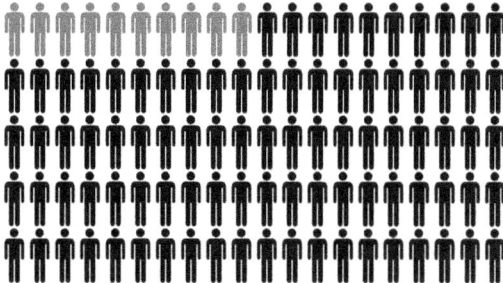

There are 100 figures representing the entire population. Ten of the figures are light gray to indicate the 10%, i.e., the 10 out of 100, who, like me, are left-handed.

11.2.3 Section Summary

In this section, we learned that percents are fractions with denominators of 100. We also learned that to convert a fraction into a percent, we first write the fraction as a decimal by dividing the numerator by the denominator, then we multiply the decimal by 100.

11.3 Percent of a Number

What is 4/5 of 20? What does this question even mean? It means: if we divide 20 into five equal-sized parts (the denominator of 5), and keep four of them (the numerator of 4), what number do we have?

Dividing gives us 20 ÷ 5 = 4. That's 1/5 of 20. To get 4/5, we multiply 1/5 of 20 by 4: 4 × 4 = 16. Therefore, 4/5 of 20 is 16. We can check that we're correct by writing 16/20 as a fraction and reducing it,

$$\frac{16}{20} = \frac{8}{10} = \frac{4}{5}$$

Here, I used the trick from Chapter 8 to reduce a fraction by first dividing numerator and denominator by 2, then 3, then 5, etc. Here, I only needed 2.

So, asking for a fraction of a number means multiplying the number by that fraction, or equivalently, by dividing the number by the denominator, then multiplying that answer by the numerator.

Now, what is 60% of 20? This is the form of the question you'll most often encounter. The question is asking for the number that covers 60/100-ths of the distance from 0 to 20 on the number line.

There are two ways to answer the question. One is to do what we did above, divide 20 by the denominator of 100 to get 0.2, then multiply that value by 60. To multiply 0.2 × 60 means to multiply 2 × 60 = 120 and move the decimal point one place to the left to get: 0.2 × 60 = 12.0. So, 60% of 20 is 12. Is it? Let's write 12/20 and reduce it,

$$\frac{12}{20} = \frac{6}{10} = \frac{3}{5}$$

well, that doesn't look like 60/100, but it is,

$$\frac{60}{100} = \frac{30}{50} = \frac{15}{25} = \frac{3}{5}$$

So, yes, 60% of 20 is 12.

11.3. PERCENT OF A NUMBER

The second way to find the answer is to write 60% as a decimal and multiply,

$$20 \times 60\% = 20 \times 0.60 = 20 \times 0.6 = 12.0$$

I suggest you pause here and review the two approaches above before continuing.

Why are the two approaches the same? The first, the one dividing the number by 100, is, in reality, calculating 1% of the number. This is because dividing by 100 is the same as multiplying by 1/100, which is the very meaning of 1%. Thus, once we have 1%, we multiply by however many percent we really want.

The second approach combines the two steps of the first approach into one. Making the percent a decimal already divides by 100, so only the multiplication part remains. Either method works, so use the one you feel most confident with.

Let's try a few more examples. What is 37.5% of 8? Let's use the second approach, that of multiplying by the percent as a decimal. To convert the percent to a decimal, move the decimal point two places to the *left*, i.e., divide by 100. This gives us 0.375. So, to get 37.5% of 8, we multiply,

$$0.375 \times 8 = 3.000 = 3$$

Sometimes, the percent we want is a "standard" one that we can immediately understand as a common fraction. For example, what is 50% of 48? Well, if all of something is 100% of it, then 50% is half of it, and half of something is dividing by 2, so 50% of 48 is $48 \div 2 = 24$. An alternative way to see this is to realize that 50% is another way to talk about 1/2 because,

$$50\% = \frac{50}{100} = \frac{25}{50} = \frac{5}{10} = \frac{1}{2}$$

Likewise, asking for 25% of something is the same as asking for 1/4 of it, so divide by 4.

We're not restricted to integers, what is 3.27% of 354.09? There is no obvious short cut here, so we convert the percent into a decimal and multiply the two numbers, 0.0327×354.09,

$$\begin{array}{r} 35409 \\ \times 327 \\ \hline 247863 \\ 70818 \\ +106227 \\ \hline 11578743 \end{array}$$

To finish, we move the decimal point six places to the left to get,

$$3.27\% \times 354.09 = 11.578743$$

Notice, I wrote the percent, not the decimal, but that's okay. Again, when you see 3.27%, think either 3.27/100 or 0.0327. The answer is messy, but no one ever said math was always clean and tidy.

Now, let's find $83.\overline{3}\%$ of 114.

I see a potential issue, $83.\overline{3}\%$ has a pesky infinitely repeating 3. How can we multiply with that? We can certainly divide by 100 to get $0.83\overline{3} = 0.8\overline{3}$, but still, the repeating 3 is there.

To answer the question, we can't multiply 114 by this number, at least, not exactly. We have two choices. One is to truncate $0.8\overline{3}$ and get an approximate answer. The other is to realize that $0.8\overline{3}$ is the decimal version of 5/6. How? By a bit of algebraic slight of hand,

$$x = 0.8\overline{3}$$
$$100x = 83.\overline{3}$$
$$10x = 8.\overline{3}$$
$$100x - 10x = 83.\overline{3} - 8.\overline{3}$$
$$90x = 75$$
$$x = \frac{75}{90} = \frac{5}{6}$$

The essence of the steps above is to set up multiples of x that, when subtracted, cancel all the repeating digits. This leaves only integers leading to $x = 0.8\overline{3} = 5/6$. We used a similar trick in Chapter 10 to show that $0.\overline{9} = 1$.

If we use the fraction, then $83.\overline{3}\%$ of 114 is,

$$\left(\frac{5}{6}\right)\left(\frac{114}{1}\right) = \frac{570}{6} = \frac{285}{3} = 95$$

Notice, the answer, 95, is exact, meaning 5/6 of 114 is exactly 95.

11.3. PERCENT OF A NUMBER

The approximate answer comes from multiplying with an approximatation of $83.\overline{3}\%$. Let's use 83.33% = 0.8333, where I'm keeping only two digits after the decimal point. In that case, we have,

$$\begin{array}{r} 8333 \\ \times 114 \\ \hline 33332 \\ 8333 \\ +8333 \\ \hline 949962 \end{array}$$

This gives us 83.33% × 114 = 94.9962, which is quite close to 95. If we had used 83.333% instead, our answer would be 94.99962, while 83.3333% would give us 94.999962. Every additional 3 in the percent adds an additional 9 to the answer. After a few iterations, we might come to believe that using an infinite number of 3's would give us 95 as the answer. Practically, however, we often need to approximate, especially when working with decimal numbers.

A final example will drive home the process. I stated above that percents greater than 100 are allowed. Therefore, what is 124% of 1125? First, what does a percent greater than 100% even mean if all of something is 100% of it? A percent greater than 100 can be thought of as the sum of two percents,

$$124\% = 100\% + 24\%$$

and, we know what 100% of 1125 is, it's all of it: 1125. Therefore, we need only find 24% of 1125 and add that number to 1125. Let's do it,

$$24\% \times 1125 = 0.24 \times 1125 = 270$$

For this example, I leave the multiplication to you. Okay, we now know that 24% of 1125 is 270. Therefore, 124% of 1125 is,

$$1125 + 270 = 1395$$

Notice, we end up with a number that is larger than the one we started with, which makes sense because 124% > 100%.

We don't need to split the percent as we did. Instead, we can proceed as before and multiply 124% × 1125, which becomes 1.24 × 1125. Again, I leave the multiplication to you. When you do it, however, you'll get 1395 as your answer. Recall, 124% = 124/100, an improper fraction greater than 1, and multiplying a number by something greater than 1 will necessarily result in a number larger than the starting number.

11.3.1 X is What Percent of Y?

Sometimes, we encounter a situation where we know a number is some percent of another, but we don't know the original number. For example, 37 is 50% of what number?

To solve problems like this, I find it helpful to think of the proportion,

$$\frac{37}{x} = \frac{50}{100}$$

On the right is the fraction representing 50%. On the left is the fraction that we need to find so that it equals 50%. We know the numerator, 37. It's the denominator that we don't know. To solve this problem, we need to find x, the number that when we divide 37 by it gives us $50/100 = 0.5 = 50\%$.

The solution is to multiply 37 by 100, then divide by 50, the percent we were given,

$$\frac{37 \times 100}{50} = \frac{3700}{50} = \frac{370}{5} = 74$$

So, 37 is 50% of 74. We can check by forming the fraction,

$$\frac{37}{74} = \frac{1}{2} = 0.5 = 50\%$$

Great, but tedious. Another approach that works best when you have a calculator handy is using the percent's decimal form. In that case, to find the answer, divide the number by the decimal form of the percent,

$$\frac{37}{0.5} = 37 \times \left(\frac{1}{0.5}\right) = 37 \times 2 = 74$$

where $1/0.5 = 2$. This is the part where the calculator helps. You might see that $0.5 + 0.5 = 1$ so that $1/0.5 = 2$, but in general, things won't be so nice.

Let's try another example, one that isn't nice. We solve the problem in precisely the same way regardless. So, 14.5 is 61.09% of what number?

The decimal form of the percent is 0.6109, so we divide 14.5 by that,

11.3. PERCENT OF A NUMBER

$$\frac{14.5}{0.6109} = 23.7354\ldots$$

Even though we don't have an exact answer, we can check if we're in the ballpark by dividing,

$$\frac{14.5}{23.7354} = 0.6109 = 61.09\%$$

This tells me that our answer to four decimals is quite good.

Let's do one more: 14 is 25% of what number? Using the proportion, we're looking to solve something like this,

$$\frac{14}{x} = \frac{25}{100} = \frac{1}{4}$$

We're looking for an x that makes the left side equal to 1/4.

We could proceed as before and divide 14 by 0.25, the percent in decimal form, or we might look at the equation above and think of the proportions. What happens if I multiply 1/4 by the 14/14 alias for 1?

$$\left(\frac{1}{4}\right)\left(\frac{14}{14}\right) = \frac{1 \times 14}{4 \times 14} = \frac{14}{56}$$

Notice, the fraction on the right has a numerator of 14 *and* is equal to 1/4 = 0.25 = 25%. Therefore, 14 is 25% of 56; we have our answer. I leave it as an exercise for you to divide 14 by 0.25. If you don't get 56, check your work.

Finally, notice that dividing 14 by 0.25 is the same as dividing by 1/4. However, we also know that dividing by a fraction is the same as multiplying by its reciprocal, so we have,

$$14 \div \frac{1}{4} = 14 \times \frac{4}{1} = 14 \times 4 = 56$$

Therefore, if you know the fraction representing the percent, you can multiply by its reciprocal and get the answer that way.

11.3.2 Section Summary

To find a percent of a number, write the percent as a decimal by dividing by 100. Then, multiply the decimal by the number. That's the central message of this section. If given a fraction instead, multiply as you usually would with fractions and reduce any answer. If the percent has a repeating decimal, your best bet, in practical terms, is to approximate your answer by keeping the first few digits of the decimal.

For cases where you are told that X is some percentage of another number you don't know, call it Y, find Y by dividing X by the percentage in decimal form, or by inspection of the proportion, which works when the percent is easily represented as a fraction, like 50% or 25%, or, if it's easier, multiply by the reciprocal of the percent as a fraction.

11.4 Percents in Action

Let's put our percent expertise to good use. In this section, we'll work through several different scenarios, or word problems, all of which involve percents. The word problems involve a bit of reasoning, but that shouldn't concern us much at this point; we've got this.

11.4.1 Recognizing Percents

Percents don't always get in our faces. Sometimes, percents are hiding, implied by the numbers around us. Consider this paragraph:

> The bustling city of Flutropolis has a population of 309,557 fluzoobles. Now, fluzoobles, the current in-vogue, three-headed pets, come in three colors: mauve, chartreuse, and plaid. Ignore for the moment that "plaid" isn't a color. There are 53,445 mauve fluzoobles in Flutropolis. Likewise, the city houses 44,007 chartreuse fluzoobles.

Approximately what percent of the fluzoobles in Flutropolis are plaid?

To answer the question, we need to know how many plaid fluzoobles there are in Flutropolis. That's straightforward as we know the number of mauve and chartreuse fluzoobles, and any fluzooble that isn't mauve or chartreuse must be plaid. Therefore, there are

$$309,557 - (53,445 + 44,007) = 309,557 - 97,452$$
$$= 212,105 \text{ plaid fluzoobles}$$

Of the 309,557 fluzoobles in Flutropolis, 212,105 of them are plaid. So, the fraction of plaid fluzoobles is,

$$\frac{212,105}{309,557}$$

11.4. PERCENTS IN ACTION

This is because we have a part, the plaid fluzoobles, out of a whole, all the fluzoobles. Fantastic! To get the percentage of plaid fluzoobles, we need the fraction above as a percent. We know how to do that; we divide 212,105 by 309,557 using long division, then multiply that answer by 100.

However, I don't want to do that; it's a pain. Notice, the question asked for the "approximate" percentage. So, we get to cheat a little. How? Well, 212,105 is close to 200,000 on the number line. Likewise, 309,557 is close to 300,000. Therefore, let's work with 200,000 and 300,000. Additionally, notice,

$$\frac{200,000}{300,000} = \left(\frac{2}{3}\right)\left(\frac{100,000}{100,000}\right) = \frac{2}{3}$$

So, in the end, we need to convert 2/3 to a percent. We know that $2/3 = 0.\overline{6}$, therefore, multiplying by 100 tells us that approximately $66.\overline{6}\%$, which we can simplify and write as 67% of the fluzoobles in Flutropolis are plaid.

What's the exact percentage of plaid fluzoobles? To get that, we need the first fraction as a decimal. And to get that, we use a calculator. Here's what I get to nine decimals,

$$\frac{212,105}{309,557} = 0.685188834\ldots$$

Notice the ... to indicate that the decimal continues. It will ultimately terminate or repeat because we're dividing two integers, but even out to thirty places, it still has not started to repeat. If you're curious, the decimal does start to repeat after 24,372 digits, but only a genuinely pedantic person would bother to discover that. Of course, I just did.

How did we do with our approximation? We said that approximately 67% of the fluzoobles are plaid. The direct calculation gives 68.52%, which is quite close, so our approximation is pretty good.

We haven't worked with examples this complex before. However, to apply what we are learning, we sometimes have no choice but to dive into the deep end of the pool and reason our way to an answer that makes sense.

Let's work on a simpler problem of this type. The process is the same, but I'll make the numbers nicer, so we don't need to approximate or use a calculator. Try this one yourself before reading my answers,

> Smitten Bob sends Alice love letters in code. Jealous
> Eve makes it her sworn duty to intercept and decode as

many of the frequently sappy missives as possible. Perhaps Eve isn't jealous so much as amused by Bob's attempts at flattery, to say nothing of poetry. Regardless, over the past month, Bob has inundated Alice's mailbox with 72 letters, or at least he attempted to. Eve successfully intercepted 18 letters and decoded 9 of them, which she posted online to Bob's unending horror.

What percent of Bob's letters did Eve intercept? What percent did Alice receive? What percent did Eve not only intercept but decode? And, finally, why hasn't Alice told Bob to knock it off already? I mean, 72 letters in a month? Get real.

Let's pull out the relevant information. The total is 72 letters, that's 100%. Eve, seemingly unconcerned by the fact that tampering with the mail is a felony, captured 18 of them and decoded 9 of those.

What percent of Bob's letters did Eve intercept?

She took 18 out of 72, or 18/72 of the letters. Reduce the fraction,

$$\frac{18}{72} = \frac{9}{36} = \frac{3}{12} = \frac{1}{4}$$

Eve took one-fourth of the letters. We know that 1/4 = 0.25 = 25%, so she intercepted 25% of the letters.

What percent did Alice receive?

If Eve captured 25%, then 100% − 25% = 75% went to Alice. Alternatively, we know Alice received 72 − 18 = 54 letters, or 54/72 of the letters. Reducing that fraction gives us,

$$\frac{54}{72} = \frac{27}{36} = \frac{9}{12} = \frac{3}{4}$$

and, 3/4 = 3 × (1/4) = 3 × 25% = 75%.

What percent did Eve not only intercept but decode?

Eve decoded 9 letters, half of the 18 letters she intercepted. If she intercepted 25%, and she decoded half, then she decoded 25% ÷ 2 = 12.5%.

As for why Alice hasn't told Bob to knock it off already, who knows? Perhaps she's just a hopeless romantic. There are worse things to be.

11.4. PERCENTS IN ACTION

11.4.2 What's The Tip?

A frequent task involves calculating the proper tip, at least in countries where tipping is expected. Consider this scenario:

> Daphne took her friend, Emma, out to dinner for her birthday. The bill was $144. Daphne thinks the server did a great job and is rather cute, so she wants to leave a 20% tip instead of the more typical 15%.

What is 20% *of* 144?

When working with "common" percents, like 20%, we have a couple of options. I'll use the two standard approaches in this answer, then show you a quicker way to get 20% below.

First, recognize that 20% = 0.20 = 20/100 = 1/5, so, 20% of something is one-fifth of it. Therefore, divide 144 by 5,

$$\begin{array}{r} 28.8 \\ 5 \overline{\smash{)}144.0} \\ \underline{10} \\ 44 \\ \underline{40} \\ 40 \\ \underline{40} \\ 0 \end{array}$$

This tells us that the tip should be $28.80.

If we don't recognize 20% = 1/5, we can still multiply by 0.2 like so,

$$\begin{array}{r} 144 \\ \times 2 \\ \hline 288 \end{array}$$

Again, telling us that the tip should be $28.80.

What is 15% *of* 144?

To find the 15% tip, we can multiply by 0.15, but we can do it faster if we realize that 15% = 10% + 5% and that 5% is half of whatever 10% is.

So, we need 10% of $144. That's easy: move the decimal point one place to the left. Therefore, 10% of $144 is $14.40. And, 5% is half of that, which, dividing by 2, is $7.20. Therefore, the 15% tip is $14.40 + $7.20 = $21.60.

The trick above works for 20% as well. Instead of adding half of 10% back in, double whatever 10% is. Therefore, we get $14.40 × 2 = $28.80, as we found above.

Finally, notice that calculating the tip, which you add back into the total cost of the meal, is the same as calculating sales tax. If the sales tax on the $144 bill was 7.5%, then we can either multiply 144 by 0.075 or calculate the 15% tip and divide that by 2 because 7.5 × 2 = 15. In this case, then, as we know 15% of $144 is $21.60, we find the sales tax to be,

$$\$21.60 \div 2 = \$10.80$$

So, Daphne's final cost for the meal, with a 15% tip and 7.5% sales tax, is,

$$\$144 + \$21.60 + \$10.80 = \$176.40$$

I hope Emma appreciates her friend.

11.4.3 Poverty Rates

One definition of poverty is earning less than $5.50 per day using 2011 USD. By that definition, according to recent World Bank data, the approximate number of people in poverty for the following countries is,

Country	Number of People in Poverty
Canada	266,070
Romania	2,121,900
Senegal	14,747,940
Serbia	1,312,520
United Kingdom	672,200
United States	5,601,500

Granted, even one person living in poverty is a failure; however, based on the numbers above, which countries have a severe problem with poverty? By the numbers, we might think that Senegal and the United States have serious problems. After all, over 14 million people in Senegal and 5.6 million in the United States live in poverty.

However, we can't meaningfully compare countries using raw numbers. Why? Because the population of these countries varies greatly. There are about 7 million people in Romania versus over 329 million in the United States. So, we might expect more people are living in poverty in the United States based solely on the much larger population. We need to remove the population difference effect between countries.

Is there anything we can do to help us meaningfully compare poverty between countries? Yes, there is. Instead of talking about absolute numbers, we should talk about *rates*. A rate, in this case, is another word for talking about percents. What percent of each country lives in poverty? Comparing percents is meaningful because the denominator for a percent is always 100, regardless of the actual population. Percents are directly comparable.

How do we get the percents? We divide the number of people living in poverty by the population of the country. This is nothing more than the usual "part of the whole" that we used previously to talk about fractions.

If we divide each number in the table above by the population of the country, and multiply that number by 100 to make it a percent, we get the following,

Country	Percent of People in Poverty
Canada	0.7%
Romania	11.0%
Senegal	88.1%
Serbia	19.0%
United Kingdom	1.0%
United States	1.7%

Now that we can compare percents, we see that the United States has a poverty rate of approximately 1.7% while Romania's poverty rate is 11.0% and Senegal is the worst off with a poverty rate of 88.1%. In Senegal, about 9 out of 10 people live in poverty. Poverty in the United States affects approximately 1 out of every 50 people. Canadians should feel proud because fewer than 1 person out of every 100, on average, lives in poverty in Canada.

11.4.4 A Tale of Two AIs

There can be no denying it; artificial intelligence, or AI, has already significantly transformed our lives and will continue to do so for the foreseeable future. So let's jump on the bandwagon and play a bit with AI and, along the way, learn how to use our newfound percent prowess to help us determine which of two AI models we might consider to be better.

For us, a *model* is a deep learning system that takes inputs, here small color images of animals and vehicles, and produces an output label: "animal" or "vehicle." The models are trained by showing them many thousands of images of animals and vehicles for which we know the proper output label. After training, the models are tested by showing them new images of animals and vehicles while keeping a tally of the label the model assigned to each known input.

I made two deep learning models of a kind known as a convolutional neural network. This is essentially the type of AI model used to label people in images, etc. The difference between the two models has to do with their internal structure. We'll just refer to them as Thing 1 and Thing 2.

I trained the models by showing them a collection of 50,000 small, 32x32 pixel, color images like these,

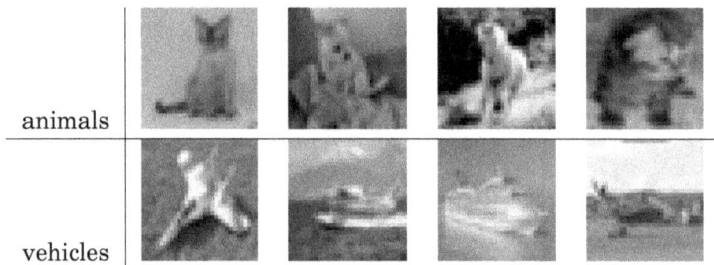

After training, I tested the two models by passing them animal and vehicle images while keeping track of the models' output. I then made two tables, one for Thing 1 (left) and the other for Thing 2 (right),

	vehicle	animal
vehicle	3820	180
animal	526	5474

	vehicle	animal
vehicle	3869	131
animal	1079	4921

Let me explain what the tables are telling us. We'll use the numbers in the tables to calculate some percentages.

Each row of the table refers to the actual label, the type of image given to the model. The columns correspond to the model's output, the label it assigned to the image. Each number in the table is a count, the number of times that possibility happened. For example,

11.4. PERCENTS IN ACTION

if the image I gave the model was of a vehicle, and the model replied with "vehicle," then the count in the upper left part of the table was increased by one. However, if the input was the image of an animal and the model replied with "vehicle," then the count in the lower left part of the table was increased by one.

Models that make no mistakes have zeros in the lower left and upper right parts of the tables because every vehicle input generates the output "vehicle," and every animal input generates "animal" as the output. Neither Thing 1 nor Thing 2 was perfect; both models made plenty of mistakes.

Okay, let's get down to some percents. One way to measure how well a model performs is to look at its *accuracy*, at the percent of inputs it correctly labeled. We can get this from the tables above. The correctly labeled inputs are those in the upper left and lower right cells of the tables. As all images end up in the table, the total number of images is the sum of all the numbers in the table.

Therefore the accuracy of Thing 1 is,

$$\frac{3820+5474}{3820+180+526+5474} = \frac{9294}{10,000} = 0.9294 = 92.94\%$$

The fraction of inputs correctly labeled by Thing 1 forms the numerator. That's the "part." The "whole" is all the inputs, the denominator. There were 10,000 test images. This is convenient because converting the fraction 9294/10,000 into a decimal means moving the decimal point four places to the left to get 0.9294. To make 0.9294 a percent, move the decimal point two places back to the right. The result is that Thing 1 was correct approximately 93% of the time because 92.94 is very close to 93 on the number line.

What about Thing 2? A similar calculation using the table on the right above gives us,

$$\frac{3869+4921}{3869+131+1079+4921} = \frac{8790}{10,000} = 0.8790 = 87.90\%$$

We can now compare the two accuracies. Which model is better? Thing 1 is better, as it was correct 93% of the time while Thing 2 was correct only 88% of the time.

What percent of the time were the models correct when they output "vehicle" for some input? We can calculate this number from the tables as well. We need the first column, the column labeled "vehicle." The number of vehicle inputs labeled "vehicle" divided by the total number of times the model output "vehicle" is what we need. Like so,

$$\text{Thing 1} = \frac{3820}{3820 + 526} = 0.8790 = 87.90\%$$
$$\text{Thing 2} = \frac{3869}{3869 + 1079} = 0.7819 = 78.19\%$$

Again, Thing 1 wins as there is an 88% chance that when it says "vehicle," it's correct. When Thing 2 says "vehicle," it's only right about 78% of the time.

Did you notice something new in the sentences above? I used the word "chance" when talking about a percent. Percents are often used to discuss probabilities, the likelihood of something happening. If something is certain to happen, the probability is 100%. Likewise, if there is no possibility of it happening, the probability is 0%. If it is equally likely to happen or not happen, then the probability is 50%, which is the origin of the phrase "50-50." We briefly explored probability in the "Think About It" section of Chapter 8.

11.4.5 Compound Interest

Here's a little story:

> Three-year-old Ashley's parents want to save for her college fund. So, they put $10,000 in the bank. The bank pays 5% interest (!) compounded annually, meaning at the end of the year, the bank adds 5% of the current balance back into the account. It's now fifteen years later, and Ashely is ready to start her college education.

How much money does Ashley have in the bank? We can make a table with the starting balance of $10,000 and how much is in the account at the end of each year when the bank pays the interest, the ending balance. See Table 11.1.

According to the table, at the end of fifteen years, the $10,000 has grown to over $20,000, all without adding another penny to the account. This is *compound interest*, and it shows the power of saving. The number in the *Interest Earned* column is 5% of the year's beginning balance. As hard as it is to do, it's worth it to save money at interest when you can.

11.4. PERCENTS IN ACTION

Year	Beginning Balance	Interest Earned	Ending Balance
1	$10,000.00	$500.00	$10,500.00
2	$10,500.00	$525.00	$11,025.00
3	$11,025.00	$551.25	$11,576.25
4	$11,576.25	$578.81	$12,155.06
5	$12,155.06	$607.75	$12,762.81
6	$12,762.81	$638.14	$13,400.95
7	$13,400.95	$670.04	$14,070.99
8	$14,070.99	$703.54	$14,774.53
9	$14,774.53	$738.72	$15,513.25
10	$15,513.25	$775.66	$16,288.91
11	$16,288.91	$814.44	$17,103.35
12	$17,103.35	$855.16	$17,958.51
13	$17,958.51	$897.92	$18,856.43
14	$18,856.43	$942.82	$19,799.25
15	$19,799.25	$989.96	$20,789.21

Table 11.1: Ashley's college fund balance by year.

11.4.6 It's On Sale!

Who doesn't love a good sale? Let's look at how percents play into sales. First, consider,

> Rupert loves his model trains. His favorite shop, *Bubba's Train Boutique*, is having a sale on locomotives. The normal price is $178.00, but, today only, all locomotives are 30% off.

How much money is Rupert saving, and what is the final cost of the locomotive?

Rupert is saving 30% of the full price, so we must multiply 178 by 30% = 0.3,

$$178 \times 30\% = 178 \times 0.3 = 53.40$$

Therefore, Rupert is saving $53.40. To find the final cost of the locomotive, we subtract,

$$178 - 53.40 = 124.60$$

So, Rupert need only pay $124.60. Rupert is a happy boy.

Notice, to find the sale price, we subtracted 30% of the full price from the full price. There is another, equivalent way to get the sale price: multiply by 100% − 30% = 70%. If we take 30% off, we have 70% left. So, why not multiply by 70% = 0.7 and be done with it? I leave working it out to you. If you don't get $124.60, check your work.

Calculating the sale price is straightforward. However, retailers sometimes put up signs like this: "Get two for the price of one!" How should we read that? Consider this scenario,

> Mr. McGee sells "antique" spoons for $50 each. How he has so many is anyone's guess. One day, he puts up a sign: "Get two spoons for the price of one!"

What is the total sale price for two spoons? What is the full price for the same two spoons? Finally, what percent of the full price is the sale price?

Two for one means the sale price for two spoons is $50. At $50 each, the full price for two spoons is $100. The fraction, the sale price over the full price, is,

$$\frac{50}{100} = 0.5 = 50\%$$

Therefore, the sign is the same as saying "Spoons 50% off!" Don't believe me? Fifty percent of $50 is half or $25. Two spoons at $25 each is $50, the same as the sale price.

Now, how about this one?

> *Barf's Burgers* is offering the following deal: choke down nine burgers, and the tenth one is free.

How much is Barf charging you per burger if you buy ten, excluding medical bills?

To get ten burgers, you need to spend 9 × $3 = $27. Therefore, each burger costs one-tenth of that, which we get by moving the decimal point one place to the left: $2.70 per burger.

Another way to think about the problem is that each burger costs 9/10 = 0.9 = 90% of the full price. As the full price is $3, the cost per burger becomes,

$$0.9 \times \$3 = \$2.70$$

as we found previously.

Here's a final case to contemplate,

11.5. PERCENT CHANGE

> Dewey, Cheatum, and Howe's *Emporium of Rare and Exotic Gifts* has far too many ancient oil lamps, which all look suspiciously identical. To move them out faster, they put up a sign: "Buy two lamps and get the third 50% off."

Casting gut feelings aside, you buy three lamps because the sale sounds too good to pass up. If the lamps cost $33 each, what percent off the total full price for three lamps are you getting?

First, what is full price for three lamps? It's 3 × $33 = $99. Next, what is the total price for two lamps at $33 each and one at $33 ÷ 2 = $16.50? Remember, if something is 50%, that's the same as multiplying by 0.5 or dividing by 2. So, we add,

$$\$33 + \$33 + \$16.50 = \$82.50$$

To find what percent of the full price is $82.50, we divide,

$$\frac{82.50}{99} = 0.8\overline{3} = 83.\overline{3}\%$$

I leave working through the division to you; it is relatively straightforward. As a percent, then, buying three lamps on sale costs you about 83% of full price, so the sale is 100% − 83% = 17% off; reasonable, but certainly not a steal and likely a fake anyway given the source.

11.4.7 Section Summary

This section presented us with multiple scenarios followed by a series of questions. By design, the answers to the questions involved percents. We learned how to calculate the percent of a group compared to an entire group, how to calculate a proper tip, sales tax, sale price, poverty rates, compound interest, and even how to compare two AI's to decide between them.

11.5 Percent Change

The basic idea behind *percent change* is that an initial value is increased or decreased by some percentage *of that value*. For example, if there is a 50% increase in bicycle theft between last year and this year, and last year there were 46 bikes stolen, then the number of bikes stolen this year is 46 plus 50% of 46 or,

$$46 + 46 \times 50\% = 46 + 23 = 65$$

Percent change always refers to two values and the difference between them as a percent of the first value.

Percent change might be negative. For example, instead of the unfortunate increase in bicycle crime, what if we noticed that the number of purses stolen last year was 120 and that the number for this year was only 90? What is the percent change in that case? It's the difference between the two numbers divided by the initial number, when written as a percent,

$$\frac{90 - 120}{120} = \frac{-30}{120} = \frac{-3}{12} = \frac{-1}{4} = -0.25 = -25\%$$

If you were the police chief for your town, you'd proudly report a 25% *decrease* in purse snatchings.

A large percent change isn't necessarily all that meaningful, however. If there were three cases of rabies in dogs in your town last year, and there were seven this year, the percent change is,

$$\frac{7-3}{3} = \frac{4}{3} = 1.333 = 133\%$$

A percent increase of 133% seems like a large number. It means a change that is the original value plus 1.33 times that original value.

Based on this result, the local newspaper might run a front-page headline screaming RABIES IN LOCAL DOGS UP 133%. Oh, the horror! However, when you realize that they are talking about 133% of 3 giving you 7, both of which are tiny numbers given the hundreds, if not thousands, of dogs in your town, you relax and take the paper to task for stirring up a panic in your neighbors who do not understand that percent changes need to refer to an actual number, which might itself be small.

Perspective is needed, however. If the percent change over last year was again 133%, but this time we're talking about the number of unemployed people in the town, which was 2300 last year, we are now talking about

$$2300 + 2300 \times 133\% = 2300 + 2300 \times 1.33 = 2300 + 3059 = 5359$$

unemployed people. That is a significant increase and likely something to be concerned about.

11.5. PERCENT CHANGE

If your small business profits are up 200% this month over last month, by how much did your profits increase? Say you made $1 last month; it's a tiny business. This month, you increased that by 200% giving you,

$$1 + 1 \times 200\% = 1 + 1 \times 2 = 1 + 2 = 3$$

meaning you tripled your profit from $1 to $3. A percent increase of 100% means double, and a percent increase of 200% means triple because you have to also add in the original amount, which is always 100%.

Here's a question: does a percent change of -120% from last year have any meaning when talking about the number of poodles in your town?

Pretend there were 500 poodles in your town last year. To get the number of poodles this year, knowing there is a -120% change, we need to calculate,

$$500 - 500 \times 120\%$$

because I said that percent change means adding or subtracting that percent of the original value. Let's work the problem through; however, I suspect you already see the issue,

$$500 - 500 \times 120\% = 500 - 500 \times 1.2 = 500 - 600 = -100$$

Not good. Somehow, we've ended up with negative 100 poodles. You can't have a percent decrease of more than 100% when talking about things because you can't take more away than there are of that thing. If talking about money, a -120% decrease might have meaning; for example, you might be $100 in debt.

Percent change often shows up when analyzing data. We live in a world of data; we can't escape it. The two examples that follow teach us how to pull percent changes out of data. The first concerns my attempt to successfully manage a simulated lemonade stand. The second deals with population growth in the Cayman Islands.

11.5.1 Lemonade Stand

Lemonade is an early educational video game developed by Apple in 1979 for their Apple II personal computer. The game simulates a simple lemonade stand and requires the player to decide, daily, based on the cost of supplies and the weather, how many glasses of lemonade to make, how many advertising signs to make, and how much to charge, in cents per glass.

The game simulates different weather scenarios and reports the outcome of each day. For example, from left to right we have sunny, hot and dry, cloudy, and the daily report:

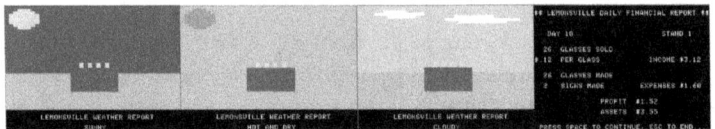

The graphics and text are retro, to be sure, but in 1979, the game was quite impressive.

The player begins with $2.00 or 200 cents. I played for ten simulated days producing the following showing the assets at the end of the day, the profit, either positive or negative, and the weather conditions,

Day	Assets	Profit	Weather
0	200		
1	200	0	sunny
2	218	18	sunny
3	248	30	sunny
4	278	30	sunny
5	238	−40	cloudy with thunderstorm
6	188	−50	sunny
7	203	15	sunny
8	355	152	hot and dry
9	365	10	sunny
10	385	20	sunny

At the end of Day 1, my profit was zero cents, meaning my expenses equaled my revenue, so my assets were still 200 cents. At the end of Day 2, however, my profit was 18 cents, giving me assets of 218 cents. My greatest profit was on Day 8, which was hot and dry. I made a profit of 152 cents and boosted my assets to 355 cents. I didn't always make money. On Day 5, it was cloudy, and a thunderstorm happened, causing me to lose 40 cents, meaning my profit was −40 cents.

Look at the outcome of Day 2, where the profit was 18 cents. At the beginning of the day, I had 200 cents in assets. At the end of the day, I had 218 cents, so my profit was 218 − 200 = 18 cents. This means the change from Day 1 to Day 2 was 18 cents out of the

11.5. PERCENT CHANGE

initial 200 cents in assets. What percent is this? It's part of the whole. Here, the whole is the 200 cents I had at the beginning of the day. Therefore, the percent change at the end of the day is,

$$\frac{18}{200} = \frac{9}{100} = 0.09 = 9\%$$

Notice, I first reduced the fraction, then, because we happened in this case to end up with a denominator of 100, I was able to write the fraction as a decimal (0.09), which, moving the decimal point two places to the right to multiply by 100, tells me that the percent change from Day 1 was a 9% increase.

In general, the percent change from an initial value at one time, let's call that value A, and the value at a later time, let's call that value B, is,

$$100\% \times \frac{B-A}{A}$$

It's important to subtract the initial value from the later value, not the other way around. If the later value is less than the initial, we want to get a negative number to understand that the percent change is negative, leading to a decrease.

Look at my Day 5 results. I started the day with 278 cents and ended with 238 cents. Let's apply the formula for percent change,

$$100\% \times \frac{238-278}{278} = 100\% \times \frac{-40}{278} = 100\% \times {}^-0.1439 = -14.39\%$$

Much is happening here. First, on the left, I'm using the percent change formula and plugging in the numbers, the ending assets for Day 5, 238 cents, and the beginning assets, 278 cents. Notice, the change is from Day 4 to Day 5, so the denominator is 278, not 238. Using the wrong denominator is a common mistake; watch for it.

The fraction −40/278 is not a convenient one to write as a decimal; it repeats after 46 digits, so I'm approximating it with four digits as −0.1439, or, as a percent, −14.49%.

Day 5 was a bad day, the thunderstorm ruined the lemonade, and I lost 14.39% of my assets. As a negative, it means the value at the later time is 14.39% less than the value at the initial time,

$$278 + 278 \times {}^-14.39\% = 278 - 40 = 238$$

To be completely honest, I'm cheating a bit on the expressions above. Multiplying 278 by 14.39% returns 40.0042, not 40, because I approximated 40/278 as a percent. However, 40.0042 is extremely close to 40, and we are only talking about whole cents, so my cheat is justified. Also, I did not compute these values by hand; I used a calculator. For the percent changes we're discussing in this section and the next, a calculator is warranted.

What happened on Day 6? It was sunny, but I still lost 50 cents. What happened is that I'm a bad manager of lemonade stands. I panicked and, in an attempt to quickly recoup my losses from the bad weather of Day 5, I made too much lemonade and it failed to sell. Poor management led to another percent change of,

$$100\% \times \frac{188-238}{238} = 100\% \times {}^{-}0.21 = -21\%$$

I lost 21% of my remaining assets on Day 6. The bad weather of Day 5 caused me to lose some 14% of my money. Poor planning on Day 6 made me lose 21%. So, in terms of effect, even though I only lost ten more cents on Day 6 than on Day 5, because I had less money to begin with, the percent change was more significant than on Day 5.

Day 8 was my best day. I made 152 cents which means a percent change of,

$$100\% \times \frac{355-203}{203} = 100\% \times \frac{152}{203} = 0.7488 = 74.88\%$$

As 74.88 is very close to 75, I'm going to claim a percent change of 75% from Day 7 to Day 8 giving me,

$$203 + 203 \times 75\% = 203 + 152 = 355$$

I ended the game after Day 10 with 385 cents in assets. I started the game with 200 cents. So, my overall percent change was,

$$100\% \times \frac{385-200}{200} = 100\% \times \frac{185}{200} = 100\% \times 0.925 = 92.5\%$$

I ended up with a percent increase of 92.5%. That's nearly 100%, meaning I came close to doubling my money. So perhaps I'm not such a lousy lemonade stand manager after all.

11.5.2 Population Growth

According to The World Bank (https://data.worldbank.org/), a graph of the population of the Cayman Islands from 1960 to 2020 looks like this,

11.5. PERCENT CHANGE

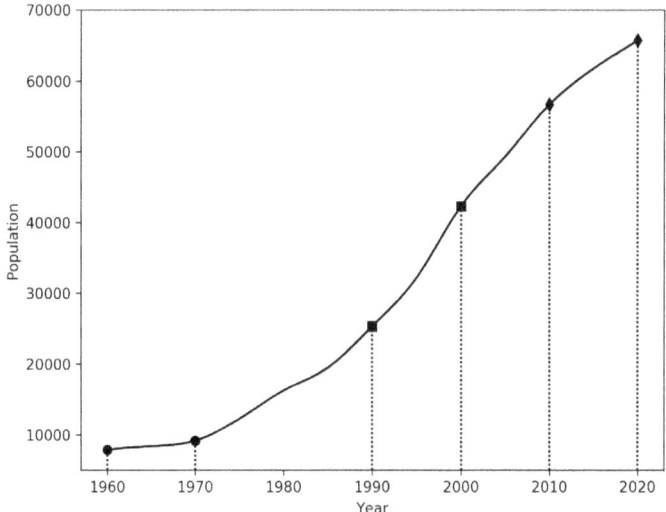

We haven't worked with graphs before, so let's walk through this one to be sure we understand what the graph is showing us. Graphs are also a frequent part of our modern lives, so we must engage with them from time to time.

This graph shows how the population of the Cayman Islands has changed between 1960 on the left and 2020 on the right. The line shows the population for each year. I put markers for specific years, along with a dashed line to make it easy to see which year I marked. We'll get to those shortly.

Between 1960 and 2020, what, overall, happened to the population of the Cayman Islands? It increased. In 1960 it was around 7500 and in 2020 grew to some 65,000 people. We can estimate these numbers by following the points marked at 1960 and 2020 to the left until they hit the vertical axis, the one labeled "Population." The number there tells us approximately what the population was. This graph is tracking a change over time. In general, graphs track how the vertical axis value, here population, changes with the horizontal axis, here time.

Okay, let's get to some percent changes. We'll calculate the percent change in population from 1960 to 1970, then from 1990 to 2000, and finally from 2010 to 2020. These are the points I marked; notice that the symbols are different to match the three time intervals. From looking at the graph, can you make any statements about these percent changes? Will they be positive or negative? Which one is likely to be the largest?

Let's begin with 1960 to 1970. From the source data, the population in 1960 was 7870 and in 1970 it was 9143. Therefore, the percent change is,

$$100\% \times \frac{9143 - 7870}{7870} = 100\% \times \frac{1273}{7870} = 100\% \times 0.1618 = 16.18\%$$

As the percent difference is positive, we know that the population increased, i.e., that

$$7870 + 16.18\% \times 7870 = 7870 + 1273 = 9143$$

What about the change in population between 1990 and 2000? The respective populations were 25,307 and 42,305 giving a percent change of,

$$100\% \times \frac{42,305 - 25,307}{25,307} = 100\% \times \frac{16,998}{25,307} = 100\% \times 0.6717 = 67.17\%$$

Again, a positive increase.

Finally, for 2010 to 2020, the respective populations were 56,672 and 65,720 for a percent change of

$$100\% \frac{65,720 - 56,672}{56,672} = 100\% \times \frac{9048}{56,672} = 100\% \times 0.1596 = 15.96\%$$

What inferences can we draw from these percent changes? The first and last are virtually identical, both 16%. The middle one, from 1990 to 2000, is much larger, 67%. Therefore, population growth in the Cayman Islands since 1960 increased more rapidly between 1990 and 2000 than either much earlier or later. On the graph, we see this as a steep rise in population between 1990 and 2000. Why the rapid increase in population? One likely factor is that since the 1970s, the Cayman Islands have become a tax haven for many companies.

11.5.3 Percentage Points

According to the American Lung Association, in 1965, about 42.4% of American adults smoked while in 2018 the number was 13.7%. Notice, these are percents of the then-current American population.

If we subtract these two numbers, we get 13.7%−42.4% = −28.7%. Does this mean there was a 28.7% decrease in smokers between

11.6. CHAPTER SUMMARY

1965 and 2018? No. When we subtract two *percents* we are calculating the *percentage points* between them, here -28.7. Calculating the *percent change* requires the formula we used above,

$$100\% \times \frac{13.7 - 42.4}{42.4} = 100\% \times \frac{-28.7}{42.4} = 100\% \times {}^-0.6768 = -67.68\%$$

Therefore, there has been a 68% decrease in smoking since 1965, not 28.7%. Good news.

Use percentage points to talk about differences between things that are already given as percents. Calculating the percent change still requires the formula.

The difference between a percentage point change and a percent change can be significant. Consider a change from 4% to 5%. The change is $5 - 4 = 1$ percentage point, seemingly small, but the actual percent change is $(5-4)/4 = 1/4 = 25\%$. A 25% change in your mortgage interest rate is a large change, not a small "1%" change.

For example, using an online mortgage calculator, a \$350,000 thirty-year loan with a fixed interest rate of 3% leads to an estimated monthly payment of \$1476. A one percentage point increase in the interest rate from 3% to 4% increases the monthly payment to \$1671. Over the thirty years of the loan, the extra paid for the "small" one percentage point change in interest amounts to \$70,200, slightly more than the median family income in the United States in 2020. Be aware of the difference between percentage point change and percent change, especially where loans are concerned.

11.5.4 Section Summary

Percent change was the focus of this section. First, we learned what a percent change is: an increase or decrease by some percentage of an initial value. We then explored two scenarios involving percent change, one managing a lemonade stand and the other interpreting population growth in the Cayman Islands. Finally, we concluded the section by learning the difference between percent change and percentage point change.

11.6 Chapter Summary

We started this chapter by answering the question, "what is a percent?" We learned that a percent is a fraction with a denominator of 100 and that percents are often given in decimal form as the fraction times 100. We learned that the percent symbol (%) is used to label a number as a percent.

We next learned how to transform fractions and decimal numbers into percents and vice versa. Following that, we learned how to find a given percent of a number by either multiplying by the fraction or by direct multiplication using the decimal version of the percent.

To understand how percents work, we explored multiple applications of percents, from calculating sales prices and tax to poverty rates between countries and compound interest, among other uses.

Finally, we learned how to calculate percent change as the change in a value from one (usually) time to another as a percent of the original value. Along the way, we explored the difference between percent change and percentage point change and learned that the two are not the same.

11.7 Terms and Concepts

We introduced the following terms and concepts in this chapter.

compound interest Interest where a percentage of an account's value is added to the balance over some time interval, for example, annually.

percent A fraction with a denominator of 100. Percents are often given as a decimal number. To convert between a percent and a decimal, divide by 100. To convert between a decimal and a percent, multiply by 100.

percentage point change The difference between two percents.

percent change The difference between two values as a percent of the first value.

rate In this chapter, a rate is synonymous with a percent. In other contexts, a rate measures how one thing changes as another changes. For example, speed is a rate; it measures how distance changes for each change in time, e.g., miles per hour.

word problem A problem presented in text form where it is necessary to pull the essential numbers from the problem to answer the questions posed.

11.8 Exercises

Exercise 1

Find the following percent of a number.

2% of 255	49% of 228	76% of 91	82% of 67
15% of 607	55% of 94	10% of 312	24% of 74
79% of 55	32% of 733	10% of 69	62% of 139
71% of 424	78% of 740	1% of 59	26% of 483
57% of 510	27% of 600	4% of 989	15% of 853

Exercise 2

It's October as I write this, Halloween season. Therefore, the following graph shows the popularity of the word "Frankenstein" from 1800 to 2020. It was created using Google's Ngram Viewer (https://books.google.com/ngrams/). The vertical scale is arbitrary; the numbers are there to show relative differences in popularity over time. Use the graph to answer the questions below. It's okay to estimate things; the graph isn't exact.

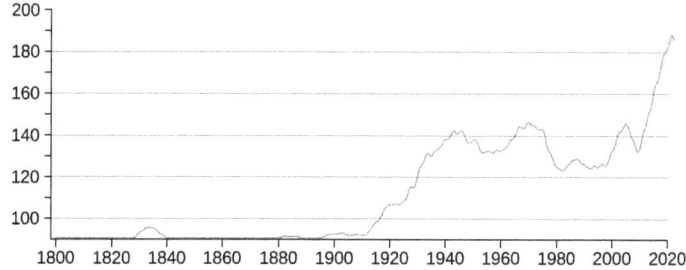

a) Frankenstein's peak popularity is around 2020. Its lowest popularity, of zero, was near 1800. What is the percent change between the peak in 2020 and 1800 when its popularity was zero? (Hint: if all of something is gone, how much of what was originally there has disappeared?)
b) What is the percent change in Frankenstein's popularity from around 1930 to 1945? Is this an increase or a decrease?
c) What is the percent change from about 1975 to 1980? Is this an increase or a decrease?
d) Why is there a blip around 1835?
e) Why might Frankenstein's popularity have suddenly increased in the 1930s and again around 2010?

Hint: Questions (d) and (e) have nothing at all to do with mathematics.

Think About It

All living things absorb carbon from their environment. This includes the most common version, or isotope, known as carbon-12 (^{12}C), which accounts for over 98% of all carbon, and the much rarer, radioactive isotope, carbon-14 (^{14}C).

Carbon-14 is created in the Earth's atmosphere by cosmic rays and nitrogen-14 (^{14}N). Nitrogen is the most abundant gas in the atmosphere. Living things absorb carbon-14 continuously until they die. At that point, the carbon-14 begins to decay at a rate where half of the carbon-14 will be gone after 5730 years. This is the *half-life* of ^{14}C.

Archaeologists have uncovered a Viking ship and want to know how old it might be. The amount of carbon-14 present in the ship's wood immediately after the ship was built is 100%. After one half-life, 5730 years, the remaining carbon-14 is 50%. After another half-life, another 5730 years or $2 \times 5730 = 11,460$ years since the ship was constructed, half of the *remaining* carbon-14 will be gone, meaning the remaining carbon-14 is $50\% \div 2 = 25\%$ of the original amount. This process continues, half-life after half-life, until all the carbon-14 atoms have decayed into nitrogen-14.

Scientists can measure the amount of carbon-14 in a sample. Based on the amount and the known half-life, they can work out the age of the sample up to a maximum age of about 50,000 years. After that point, too little carbon-14 remains to be reliably detected.

11.8. EXERCISES

Let's make a table of the percent of ^{14}C remaining from one half-life to the next,

Half-life	Time (years)	^{14}C remaining (%)
0	0	100
1	5730	50
2	11,460	25
3	17,190	12.5
4	22,920	6.25

After four half-lives, only 6.25% of the original ^{14}C remains. We can plot the percent of the original carbon-14 for ten half-lives,

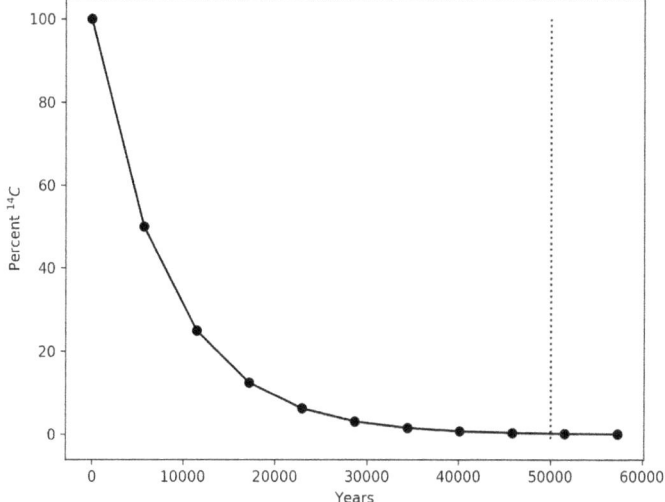

The vertical dashed line is at 50,000 to mark the ^{14}C age limit. Notice how quickly the amount decreases. A graph that looks like this is showing *exponential decay*. It's the opposite of exponential growth, which we explored briefly in Chapter 9.

Because of the 50,000 year limit, carbon-14 dating is generally not used to date fossils. However, many other radioactive isotopes can be used to date fossils or, more typically, volcanic deposits associated with the fossils. One such process is *potassium-argon* dating where potassium (^{40}K) decays into argon (^{40}Ar) with a half-life of about 1.3 billion years, far longer than the half-life of ^{14}C.

Make a table like the one above using potassium-argon dating. The table will look the same; only the time column will change. It might be helpful to note that 1.3 billion is 1300 million and work in millions of years instead. The Earth is 4.54 billion (4540 million) years old. Approximately how many potassium-argon half-lives is that?

Chapter 12

Epilogue, Resources, And Cheat Sheet

Every book should have an epilogue. You'll find one here. I also include a list of resources and an arithmetic cheat sheet; a summary of the mechanics we learned in the previous chapters.

12.1 Epilogue

"There is no real ending. It's just the place where you stop the story."
– Frank Herbert

Our story stops here. However, I sincerely hope *your* arithmetic story is just beginning. No book can be all that is needed on any topic. If that were so, there would be only one book per topic, and how boring would that be?

Therefore, seek out other books on arithmetic, and popular math, and algebra, and trig, and statistics, and everything else the fantastic world of mathematics has to offer. Arithmetic is the foundation, the bedrock. Now build something on it. Below are some resources to help you on your journey.

12.2 Resources

Below you'll find a few places you can go to further your understanding of arithmetic and math in general. I split the resources into three groups: arithmetic, arithmetic with algebra (and other topics), and what I'm calling "statistical literacy," by which I mean learning how to interpret charts and graphs combined with basic concepts behind understanding data (statistics).

Arithmetic

- *Arithmetic* by Paul Lockhart. This is a fun book. It's a casual perusal of arithmetic and focuses on thinking and fun instead of specific problems. For an alternate take on how to approach our subject, I recommend this book.

- *Faster Arithmetic for Adults* by Ken Ward. Throughout this book, I occasionally referred to shortcuts when I found them particularly easy to follow. There are many more shortcuts to arithmetic. To learn some of them, consider this book.

- *Arithmetic Refresher: Improve Your Working Knowledge* by A. A. Klaf. This is an older book, but it gets right to the heart of the practical application of arithmetic. The question and answer approach complements this book and will likely resonate with many readers.

- Khan Academy Arithmetic videos (`https://www.khanacademy.org/math/arithmetic`). Khan Academy more or less wrote the website on free online courses. The videos here cover much of the same material as this book, but, of course, as videos, not static pages.

Arithmetic with Algebra

- *Math Refresher for Adults: The Perfect Solution* by Richard W. Fisher. Fisher's book is similar to ours but a little more granular in its presentation and expanded in its topics.

- *Maths Made Easy* by Dexter J. Booth. I selected this book because, while it goes quickly through the topics we covered here, it then moves to other algebra topics. The book is also free for the price of an Internet Library account, see `https://archive.org/details/mathsmadeeasy0000boot`.

Statistical Literacy

- *How Charts Lie: Getting Smarter About Visual Information* by Alberto Cairo. We are exposed to charts and graphs constantly. We explored a few in this book. However, it's not always easy to pull meaning from charts and graphs, especially when there are straightforward ways to manipulate them to "lie" about the information presented. In some ways, *How Charts Lie* is an updated version of the book below. I recommend both.

- *How To Lie With Statistics* by Darrell Huff. This book is fun, entertaining, doubly so because it is quite dated, and the dollar amounts in his examples are quaint. If only prices were still so low!

- *Statistics Without Tears* by Derek Rowntree. Statistics is the art of extracting meaningful information from data. Rowntree's book teaches statistics without a single equation, only description, and common sense. Start here if you are completely unfamiliar with statistics.

12.3 Arithmetic Cheat Sheet

Here you'll find a cheat sheet of sorts, a summary of the mechanics related to all the operations and types of numbers that we've explored in this book.

Place Notation

Place names:

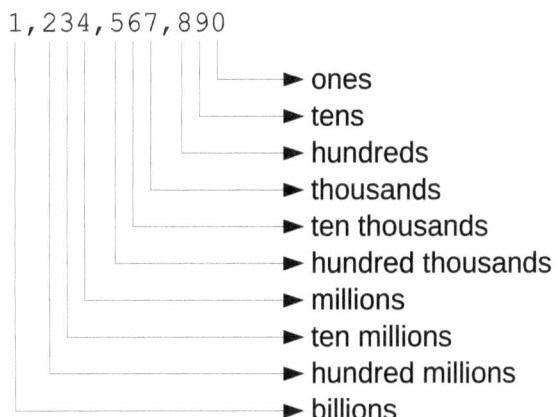

Powers of 10:

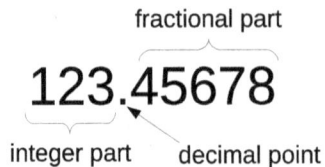

in expanded form is

$$1\times 10^2+2\times 10^1+3\times 10^0+4\times 10^{-1}+5\times 10^{-2}+6\times 10^{-3}+7\times 10^{-4}+8\times 10^{-5}$$

where each place's value is an increasing power of 10 when moving to the left and a decreasing power when moving to the right.

Addition and Subtraction

Single-Digit Addition

+	0	1	2	3	4	5	6	7	8	9
0	0	1	2	3	4	5	6	7	8	9
1	1	2	3	4	5	6	7	8	9	10
2	2	3	4	5	6	7	8	9	10	11
3	3	4	5	6	7	8	9	10	11	12
4	4	5	6	7	8	9	10	11	12	13
5	5	6	7	8	9	10	11	12	13	14
6	6	7	8	9	10	11	12	13	14	15
7	7	8	9	10	11	12	13	14	15	16
8	8	9	10	11	12	13	14	15	16	17
9	9	10	11	12	13	14	15	16	17	18

Single-Digit Subtraction

-	0	1	2	3	4	5	6	7	8	9
0	0	-1	-2	-3	-4	-5	-6	-7	-8	-9
1	1	0	-1	-2	-3	-4	-5	-6	-7	-8
2	2	1	0	-1	-2	-3	-4	-5	-6	-7
3	3	2	1	0	-1	-2	-3	-4	-5	-6
4	4	3	2	1	0	-1	-2	-3	-4	-5
5	5	4	3	2	1	0	-1	-2	-3	-4
6	6	5	4	3	2	1	0	-1	-2	-3
7	7	6	5	4	3	2	1	0	-1	-2
8	8	7	6	5	4	3	2	1	0	-1
9	9	8	7	6	5	4	3	2	1	0

Adding and Subtracting Integers and Decimals

To add two positive integers, write them vertically, lining up the ones columns. Then add place by place from right to left. If the sum exceeds 10, write the ones value of the sum in the current place and carry the 1 to the next column to the left. For example,

$$\begin{array}{r} 1111 \\ 254567599 \\ +3435 \\ \hline \mathbf{254571034} \end{array}$$

Line up the ones column when subtracting two positive integers,

$$\begin{array}{r} 8954 \\ -1723 \\ \hline \mathbf{7231} \end{array}$$

or, with borrowing,

$$\begin{array}{rl} 0\,12 & \leftarrow \text{third borrow} \\ 211 & \leftarrow \text{second borrow} \\ \cancel{1}14 & \leftarrow \text{first borrow} \\ \cancel{1}\,\cancel{3}\,\cancel{2}\,4 & \\ -5\,7\,8 & \\ \hline \mathbf{7\,4\,6} & \end{array}$$

If necessary, borrow from multiple places to the left,

$$\begin{array}{r} 3\ 9\ 9\ 9\ 9\ 11 \\ 4\ \cancel{0}\ \cancel{0}\ \cancel{0}\ \cancel{0}\ \cancel{1} \\ -\ 1\ 9\ 8\ 7\ 6\ 5 \\ \hline \mathbf{2\ 0\ 1\ 2\ 3\ 6} \end{array}$$

To add or subtract with a decimal number, line up the decimal points, copy the decimal point to the answer, then proceed as above ignoring the decimal point. For example,

$$\begin{array}{r} 1 \\ 83.7 \\ +12.88 \\ \hline \mathbf{96.58} \end{array}$$

and,

$$\begin{array}{r} 64.99 \\ -32.27 \\ \hline \mathbf{32.72} \end{array}$$

Multiplication

Single-Digit Multiplication

×	0	1	2	3	4	5	6	7	8	9
0	0	0	0	0	0	0	0	0	0	0
1	0	1	2	3	4	5	6	7	8	9
2	0	2	4	6	8	10	12	14	16	18
3	0	3	6	9	12	15	18	21	24	27
4	0	4	8	12	16	20	24	28	32	36
5	0	5	10	15	20	25	30	35	40	45
6	0	6	12	18	24	30	36	42	48	54
7	0	7	14	21	28	35	42	49	56	63
8	0	8	16	24	32	40	48	56	64	72
9	0	9	18	27	36	45	54	63	72	81

which can be reduced to the following smaller table,

×	0	1	2	3	4	5	6	7	8	9
0										
1										
2			4							
3			6	9						
4			8	12	16					
5			10	15	20	25				
6			12	18	24	30	36			
7			14	21	28	35	42	49		
8			16	24	32	40	48	56	64	
9										

if the following facts are remembered:

- Any number multiplied by zero is zero.
- Any number multiplied by one is the number itself.
- If the first number is smaller than the second, flip the order and multiply the second by the first.
- Nine times any number is the number minus one followed by whatever you need to add to that number to get nine.

Multiplying Integers and Decimals

Multiply two integers by lining up the ones columns, then multiplying the top number by each digit of the bottom, from left to right, writing the partial product beneath the current digit place. Then add all the partial products. For example,

$$
\begin{array}{r}
1\ 4\ 1\ 5\ 1\ 2\ 2 \\
1\ 6\ 1\ 8\ 2\ 3\ 4 \\
\times 7 \\
\hline
1\ 1\ 3\ 2\ 7\ 6\ 3\ 8
\end{array}
$$

for multiple digits becomes,

$$
\begin{array}{r}
1\ 1 \\
2\ 3\ 1 \\
5\ 5\ 2 \\
2\ 5\ 6\ 3 \\
\times 3\ 5\ 9 \\
\hline
2\ 3\ 0\ 6\ 7 \\
1\ 2\ 8\ 1\ 5 \\
+\ 7\ 6\ 8\ 9 \\
\hline
9\ 2\ 0\ 1\ 1\ 7
\end{array}
$$

where the three partial products are added to find the answer.

For decimals, line up the rightmost digit of each number ignoring the decimal point, then multiply. For example, 16.183×4.1 becomes,

$$
\begin{array}{r}
16183 \\
\times 41 \\
\hline
16183 \\
+64732 \\
\hline
663503
\end{array}
$$

To finish, move the decimal point to the left as many places as there are digits in the fractional part of each number. For the example here, there are three in the first number and one in the second, so move the decimal point four places to the left to get,

$$16.183 \times 4.1 = 66.3503$$

Division

The recipe for dividing a number by a single digit is:

1. Work left to right considering each digit of the dividend in turn. If the first digit of the dividend is smaller than the divisor, move to the second and divide the first two digits together.

2. Ask the question: how many X's go into Y? Where X is the divisor and Y is the current digit of the dividend. The answer that gets closest to Y is the corresponding digit of the quotient.

3. Multiply the quotient digit and the divisor, then subtract that product from the current dividend digit.

4. Bring down the next digit of the dividend.

5. Repeat from Step 2 until all the digits of the dividend have been examined. The quotient is above the dividend, and any final remainder is below in the work area as what's leftover after the final subtraction.

For example, $97,804 \div 9$ is,

$$
\begin{array}{r}
10867 \\
9\overline{)97804} \\
\underline{9} \\
078 \\
\underline{72} \\
60 \\
\underline{54} \\
64 \\
\underline{63} \\
1
\end{array}
$$

Notice, there is a remainder of 1. The remainder is always less than the divisor, and forms a fractional part. Therefore, the full answer is,

$$97,804 \div 9 = 10,867\frac{1}{9}$$

when written as a mixed number.

Long division with arbitrary divisors follows much the same process. Begin with as much of the dividend as is needed to fit at least one divisor. Consider,

$$
\begin{array}{r}
1506 \\
251\overline{)378014} \\
\underline{251} \\
1270 \\
\underline{1255} \\
1514 \\
\underline{1506} \\
8
\end{array}
$$

where it was necessary to begin with 378, then proceed by bringing down the remaining digits of the dividend one by one. Therefore,

$$378{,}014 \div 251 = 1506\frac{8}{251}$$

when writing the remainder and divisor as a fraction.

Fractions

Reducing Fractions

To reduce a fraction, divide the numerator and denominator by the greatest common divisor (GCD), the largest number that evenly divides both numbers. In practice, it's often much faster to repeatedly divide the numerator and denominator by increasing prime numbers as many times as possible before trying the next largest prime. The first few prime numbers are,

$$2, 3, 5, 7, 11, 13, 17, 19$$

For example,

$$\frac{840}{1560} = \frac{420}{780} = \frac{210}{390} = \frac{105}{195} = \frac{35}{65} = \frac{7}{13}$$

by dividing by 2 three times, 3 once, and 5 once. Dividing by 10 initially saves a few steps,

$$\frac{840}{1560} = \frac{84}{156} = \frac{42}{78} = \frac{21}{39} = \frac{7}{13}$$

where dividing by 10 is followed by 2 twice and 3 once.

Adding and Subtracting

If the fractions have a common denominator, add or subtract the numerators,

$$\frac{3}{12} + \frac{7}{12} = \frac{10}{12} = \frac{5}{6}$$

and,

$$\frac{11}{13} - \frac{6}{13} = \frac{5}{13}$$

If the denominators are not the same, find the least common multiple of each denominator, then multiply each fraction by the alias for 1 leading to that common multiple. In practice, it's often easier to multiply the first fraction by the denominator of the second and the second by the denominator of the first. Then add or subtract and reduce if necessary. For example,

$$\frac{5}{7} + \frac{1}{6} = \left(\frac{5}{7}\right)\left(\frac{6}{6}\right) + \left(\frac{1}{6}\right)\left(\frac{7}{7}\right)$$
$$= \frac{30}{42} + \frac{7}{42}$$
$$= \frac{37}{42}$$

likewise,

$$\frac{3}{4} - \frac{2}{3} = \left(\frac{3}{4}\right)\left(\frac{3}{3}\right) - \left(\frac{2}{3}\right)\left(\frac{4}{4}\right)$$
$$= \frac{9}{12} - \frac{8}{12}$$
$$= \frac{1}{12}$$

Multiplication and Division

To multiply two fractions, multiply the numerators, multiply the denominators, and reduce the answer. For example,

$$\frac{4}{7} \times \frac{8}{14} = \frac{4 \times 8}{7 \times 14} = \frac{32}{98} = \frac{16}{49}$$

To divide, flip the numerator and denominator of the second fraction, the divisor, and then multiply,

$$\frac{4}{7} \div \frac{8}{14} = \frac{4}{7} \times \frac{14}{8} = \frac{4 \times 14}{7 \times 8} = \frac{56}{56} = 1$$

Percents

Percent of a Number

To get a percent of a number, multiply the number by the percent in decimal form. For example, 46% of 138 is,

$$46\% \times 138 = 0.46 \times 138 = 63.48$$

12.3. ARITHMETIC CHEAT SHEET

If told that some number is a given percent of another, multiply the number by 100, then divide by the percent. Therefore, if told that 63.48 is 46% of a number, find the number with,

$$63.48 \times 100 = 6348 \rightarrow 6348 \div 46 = 138$$

Likewise, 12 is 32% of what number? Multiply by 100 and divide by 32,

$$12 \times 100 = 1200 \rightarrow 1200 \div 32 = 37.5$$

Percent Change

The percent change from one value to another is the difference expressed as a percent of the first value. To find the percent change from A to B, subtract A from B, divide by A, and multiply by 100,

$$\text{percent change} = 100 \times \frac{B-A}{A}$$

For example, the percent change from 24 to 36 is,

$$100 \times \frac{36-24}{24} = 100 \times \frac{12}{24} = 100 \times 0.5 = 50\%$$

The percent change is negative if there was a decrease in the value. The percent change from 22 to 16 is,

$$100 \times \frac{16-22}{22} = 100 \times \frac{-6}{22} = 100 \times \frac{-3}{11} = 100 \times {}^-0.\overline{27} = -27.\overline{27}\%$$

or a percent change of about -27.3%.

Powers

To raise a number, a, the base, to the n-th power (n the exponent), multiply a by itself n times,

$$a^n = \overbrace{a \times a \times a \times \ldots \times a}^{n}$$

Therefore, 3^4 is,

$$3^4 = 3 \times 3 \times 3 \times 3 = 81$$

When multiplying powers of the same base, the exponents add,

$$(3^3)(3^2) = 3^{3+2} = 3^5 = 243$$

When dividing powers of the same base, subtract the exponents,

$$3^6/3^4 = 3^{6-4} = 3^2 = 9$$

Negative exponents follow this formula,

$$a^{-n} = \frac{1}{a^n}$$

So that,

$$3^{-2} = \frac{1}{3^2} = \frac{1}{9} = 0.\overline{1}$$

Working with Negative Numbers

Addition and Subtraction

It's best to think your way through what makes sense when adding or subtracting positive and negative numbers. With that said, the following summary helps in determining what to do:

Addition:

pos + pos	Add normally.
pos + neg	Subtract the second from the first.
neg + pos	Subtract the first from the second.
neg + neg	Add ignoring signs. Negate answer.

Subtraction:

pos − pos	Subtract normally.
pos − neg	Add ignoring sign of the second.
neg − pos	Add ignoring signs. Negate answer.
neg − neg	Subtract first from second ignoring signs.

Multiplication and Division

For multiplication and division, the rules are straightforward:

- If *either* of the numbers is negative, the answer is *negative*.
- If *both* of the numbers are negative, the answer is *positive*.

"That's all there is, there isn't any more."
− Ethel Barrymore

Chapter 13

Solutions to Exercises

Here you will find solutions to the practice exercises at the end of each chapter. Worksheets with more practice problems are available on the book's website:

`https://github.com/rkneusel9/ArithmeticExplained`

Chapter 1

Exercise 1

- seven hundred eighty-three
- ten thousand fifty-six
- one hundred eleven thousand one hundred eleven
- one million thirty thousand five
- one hundred twenty-three million four hundred fifty-six thousand seven hundred eighty-nine

Exercise 2

```
        17
     8,087
    42,364
10,010,010
299,792,458
```

Exercise 3

$1 < 4$, **true**	$0 < 5$, **true**	$-5 < 5$, **true**
$3 > 2$, **true**	$-5 < 0$, **true**	$5 > -5$, **true**
$-4 > -3$, **false**	$-1 > 1$, **false**	$-2 > -4$, **true**

Challenge 1

42	**XLII**
127	**CXXVII**
1066	**MLXVI**
1941	**MCMXLI**
2000	**MM**
2022	**MMXXII**

Think About It

Counting with only three fingers on each hand implies that we would count with sets of six, not ten. If we use the same digits as we do for normal counting, we'd only need 0, 1, 2, 3, 4, and 5 to number the fingers on both hands as 1, 2, 3, 4, 5, 10. In this case, "10" no longer means one ten and no ones, but one *six* and no ones. So, the first ten numbers written in the place notation by people with only three fingers on each hand become 1, 2, 3, 4, 5, 10, 11, 12, 13, 14. The final number, "14," is one six and four ones, which is ten.

Far from being a strange exercise, grouping the digits by sets other than ten is commonly done, especially when working with computers that group digits by twos, eights, and sixteens. We'll investigate this more later in the book, but it's good to think about it a bit now.

305

Chapter 2

Exercise 1

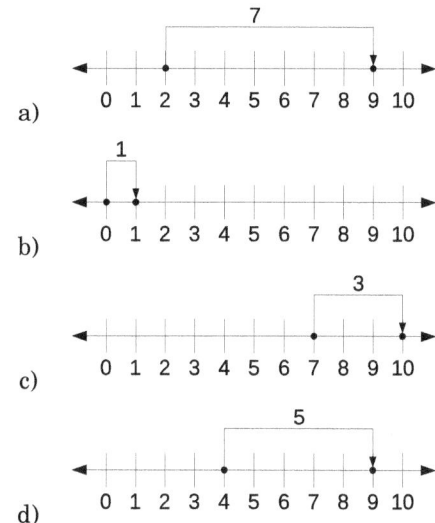

a)
b)
c)
d)

Exercise 2

```
   1380          3216          3907          3303
+  3555       +  3147       +   319       +   578
   ----          ----          ----          ----
   4935          6363          4226          3881


    287          2676           859          3594
+  1535       +  3093       +  1502       +   101
   ----          ----          ----          ----
   1822          5769          2361          3695


   2012          2333           715          3507
+  1774       +  2154       +   377       +   997
   ----          ----          ----          ----
   3786          4487          1092          4504


    699           941           343          4012
+  3387       +  3144       +  2185       +   244
   ----          ----          ----          ----
   4086          4085          2528          4256
```

Think About It

When the sum of a column exceeds 99, we have two options. The simplest option is to put the ones digit of the sum in the ones column of the answer as before, and move the remaining digits of the sum to the next column on the left, as we did before.

In this case, the sum of the ones digits is 128. So, we put 8 in the ones column of the answer and use 12 as the carry. Then, we proceed with the tens column as before.

Alternatively, we can split the value between the next column to the left, and the one after that since the sum includes a hundreds value. This requires *two* carries. Using this approach, the 128 is split into 8 in the ones column of the answer, 2 becomes the carry of the tens column, and 1 the carry of the hundreds column. Then add the tens column and hundreds column as before. If adding the tens column produces its own carry, that's fine; add it into the hundreds column along with the 1 from summing the ones column.

The second approach works because 12 tens are 120, which is one hundred and two tens. All things considered, however, it's easier to use 12 as the carry in the tens column, but it doesn't hurt to know why that's a perfectly valid thing to do.

Chapter 3

Exercise 1

```
    2989         6481         9113         5463
  - 2607       - 4211       - 4013       - 5143
  ------       ------       ------       ------
     382         2270         5100          320


    1348         2739         5343         4572
  -  233       -   35       - 4213       - 4071
  ------       ------       ------       ------
    1115         2704         1130          501
```

```
   6135        8936        5198        3591
−    33     − 2332     − 2166     − 3550
   6102        6604        3032          41


   1865        4689        6752        1343
− 1341      − 2058      − 6421      − 1310
    524        2631         331          33
```

Exercise 2

```
   4558        4984        1778        3622
− 4211      − 2528      −  765      −  665
    347        2456        1013        2957


   2904        2593        4934        1161
− 1800      −  838      − 1899      −  508
   1104        1755        3035         653


   3113        1606        2847        3468
− 1582      −  990      −  228      −  751
   1531         616        2619        2717


   3619        3649        4852        3740
− 2552      − 3370      − 2446      − 2295
   1067         279        2406        1445
```

Exercise 3

```
   10000          404004
−    499       −  39939
    9501          364065


    2001         9001900
−   1002       −   99999
     999         8901901
```

Exercise 4

Only one number satisfies each of the first four equations,

$$12 - 5 = 7$$
$$18 - 7 = 11$$
$$17 - 2 = 15$$
$$7 - 4 = 3$$

The single digit solutions to the final equation are all those where the second number is two less than the first,

$$2 - 0 = 2$$
$$3 - 1 = 2$$
$$4 - 2 = 2$$
$$5 - 3 = 2$$
$$6 - 4 = 2$$
$$7 - 5 = 2$$
$$8 - 6 = 2$$
$$9 - 7 = 2$$

In general, there are an infinite number of solutions to $\Box - \Box = 2$ because any first number two greater than the second number satisfies the equation.

Think About It

Yes, the statement is true for all numbers. Subtraction is the same as adding the opposite of the second number, see Chapter 4.

Chapter 4

Exercise 1

```
   -455         576         914        -352
+   210     + -684     +   -56     + -420
  ----        ----        ----        ----
  -245        -108         858        -772


  -537        -684        -805         673
+  700     +  -528    +   781     +   228
  ----        -----       ----        ----
   163       -1212         -24         901


   681          23        -380        -803
+  818     +   649     +  -314     +  -416
  ----        ----        ----        -----
  1499         672        -694       -1219


   829        -108         790        -852
+  756     +  -181     +   778     +  -706
  ----        ----        ----        -----
  1585        -289        1568       -1558
```

Exercise 2

```
   832         448        -362         538
-  279      -  833     -  -300     -  458
  ----        ----        ----        ----
   553        -385         -62          80


   401         879        -905         507
-  235      -  455     -    31     -  907
  ----        ----        ----        ----
   166         424        -936        -400


  -734        -257         685         340
-  616     -  -481     -  -572     -  389
  -----       ----        ----        ----
 -1350         224        1257         -49
```

```
    -283            -529            130            159
 -  -355         -  -943         -  -746        -   268
    ────            ────           ────           ────
     72             414            876           -109
```

Think About It

"Is subtraction real in the same sense that addition is real?"

There are different ways to interpret this question. Without diving into the larger question of whether math is "real" or not, or if we discover mathematics or invent it, we might narrowly read this question as asking if there is any difference between how we might think of subtraction compared to addition.

In Chapter 4 we learned that $8 - 3 = 8 + {}^-3 = 5$, i.e., we learned that subtraction can be interpreted as adding a negative number. In that sense, we might say "no," subtraction is not real in the sense that addition is real.

Mathematicians often act this way. They talk about addition without mentioning subtraction because addition is all you need. Also, adding the opposite of a number has a special meaning in mathematics. For example,

$$12 + {}^-12 = 0$$

Mathematicians would say that -12 is the *additive inverse* of 12 because when you add the two numbers together, you get zero. Every number has an additive inverse and, for the numbers we're working with, that additive inverse is what you get when you change the sign. So, the additive inverse of 42 is -42 and the additive inverse of -67 is 67.

Chapter 5

Exercise 1

```
     6           2           9           4
 ×   5       ×   9       ×   8       ×   4
 ─────       ─────       ─────       ─────
    30          18          72          16

     4           5           6           7
 ×   7       ×   2       ×   3       ×   9
 ─────       ─────       ─────       ─────
    28          10          18          63

     8           3           5           8
 ×   9       ×   2       ×   7       ×   8
 ─────       ─────       ─────       ─────
    72           6          35          64

     3           3           7           4
 ×   5       ×   7       ×   6       ×   8
 ─────       ─────       ─────       ─────
    15          21          42          32

     7           2           9           9
 ×   2       ×   4       ×   4       ×   5
 ─────       ─────       ─────       ─────
    14           8          36          45
```

Exercise 2

```
      34          81          17          38
  ×   79      ×   89      ×   66      ×   36
  ──────      ──────      ──────      ──────
     306         729         102         228
  + 2380      + 6480      + 1020      + 1140
  ──────      ──────      ──────      ──────
    2686        7209        1122        1368

      57          60          59          24
  ×   41      ×   41      ×   19      ×   55
  ──────      ──────      ──────      ──────
      57          60         531         120
  + 2280      + 2400      +  590      + 1200
  ──────      ──────      ──────      ──────
    2337        2460        1121        1320
```

312   CHAPTER 13. SOLUTIONS TO EXERCISES

```
        61              76              36              26
  ×     11        ×     18        ×     14        ×     35
        61             608             144             130
  +    610        +    760        +    360        +    780
       671            1368             504             910

        28              63              31              38
  ×     39        ×     89        ×     56        ×     95
       252             567             186             190
  +    840        +   5040        +   1550        +   3420
      1092            5607            1736            3610
```

Exercise 3

```
       798             385             180             363
  ×    555        ×    642        ×    391        ×    312
      3990             770             180             726
     39900           15400           16200            3630
  + 399000        + 231000        +  54000        + 108900
    442890          247170           70380          113256

       828             829             510             298
  ×    179        ×    339        ×    945        ×    175
      7452            7461            2550            1490
     57960           24870           20400           20860
  +  82800        + 248700        + 459000        +  29800
     148212          281031          481950           52150

       312             595             842             421
  ×    618        ×    686        ×    135        ×    522
      2496            3570            4210             842
      3120           47600           25260            8420
  + 187200        + 357000        +  84200        + 210500
    192816          408170          113670          219762
```

```
            641              547              834              891
      ×     742        ×     646        ×     137        ×     816
           1282             3282             5838             5346
          25640            21880            25020             8910
      + 448700        + 328200        +   83400        + 712800
          475622           353362           114258           727056
```

Exercise 4

1. negative
2. negative
3. positive
4. negative
5. positive
6. neither, zero is not positive or negative

Think About It

Two observations explain why the process works. The first observation has to do with multiplying the numbers on the left column by two each time. The sequence of numbers produced, the powers of two, are special. They represent the place notation values for a number system based on 2, not 10. This number system is called *binary*; it's what computers use internally. Any integer can be expressed in binary, so the scribes could always find a combination of numbers on the left that add up to the multiplier.

The second observation is that the sequence of numbers on the right is the first number times the corresponding power of two. For the example,

$$2 \times 82 = 164$$
$$4 \times 82 = 328$$
$$8 \times 82 = 656$$
$$16 \times 82 = 1312$$
$$32 \times 82 = 2624$$

If we add all the equations above for the multipliers that sum to 44 we can write,

$$82 \times 2 + 82 \times 4 + 82 \times 8 + 82 \times 32$$

Doing this is the same as adding the marked numbers on the right, except that I'm replacing each number with the product that produced it. To see why this is 82 × 44, I need to use a little algebra that we haven't learned yet. The algebra I need is called the *distributive property*, and it says that if I'm adding a bunch of things each multiplied by the same thing, like 82, that's the same as adding the bunch of things first, then multiplying the sum by 82. So, the expression above becomes,

$$82 \times (2 + 4 + 8 + 32) = 82 \times 44$$

which is precisely what we wanted to multiply. The numbers added together in the parentheses, which means add them first, are precisely the powers of two the scribe would have marked.

Chapter 6

Exercise 1

$73 \div 31 = 2r11$ $168 \div 46 = 3r30$ $151 \div 37 = 4r3$
$67 \div 42 = 1r25$ $263 \div 48 = 5r23$ $210 \div 26 = 8r2$
$193 \div 25 = 7r18$ $171 \div 41 = 4r7$ $268 \div 43 = 6r10$

Exercise 2

```
         9508              3399              7045             3610
      4)38034           6)20397           4)28182          2)7221
        36000             18000             28000            6000
        -----             -----             -----            ----
         2034              2397               182             1221
         2000              1800               160             1200
         ----              ----               ---             ----
           34               597                22               21
           32               540                20               20
           --               ---                --               --
            2                57                 2                1
                            54
                            --
                             3

         7290              4758             43389            23087
      7)51030           3)14275           2)86778          4)92351
        49000             12000             80000            80000
        -----             -----             -----            -----
         2030              2275              6778            12351
         1400              2100              6000            12000
         ----              ----              ----            -----
          630               175               778              351
          630               150               600              320
          ---               ---               ---              ---
            0                25               178               31
                             24               160               28
                             --               ---               --
                              1                18                3
                                               18
                                               --
                                                0

         2583             11406             16605             3465
      9)23252           3)34219           4)66420          4)13860
        18000             30000             40000            12000
        -----             -----             -----            -----
         5252              4219             26420             1860
         4500              3000             24000             1600
         ----              ----             -----             ----
          752              1219              2420              260
          720              1200              2400              240
          ---              ----              ----              ---
           32                19                20               20
           27                18                20               20
           --                --                --               --
            5                 1                 0                0
```

$$\begin{array}{r}6250\\9\overline{)56253}\\54000\\\hline 2253\\1800\\\hline 453\\450\\\hline 3\end{array} \qquad \begin{array}{r}992\\4\overline{)3971}\\3600\\\hline 371\\360\\\hline 11\\8\\\hline 3\end{array} \qquad \begin{array}{r}28712\\2\overline{)57424}\\40000\\\hline 17424\\16000\\\hline 1424\\1400\\\hline 24\\20\\\hline 4\\4\\\hline 0\end{array} \qquad \begin{array}{r}9904\\7\overline{)69334}\\63000\\\hline 6334\\6300\\\hline 34\\28\\\hline 6\end{array}$$

Think About It

Let's work through the process for $151 \div 4$ and see if we can't notice where the remainder appears. First, we need the table,

1	4
2	8
4	16
8	32
16	64
32	128

Then, we start with the second column and work upward adding values to our running total as long as adding those values does not exceed 151. Doing this gives,

1	4	←	keep, $144 + 4 = 148$
~~2~~	~~8~~		ignore, $144 + 8 > 148$
4	16	←	keep, $128 + 16 = 144$
~~8~~	~~32~~		ignore, $128 + 32 > 148$
~~16~~	~~64~~		ignore, $128 + 64 > 148$
32	128	←	initial sum, 128

The table is identical to the table in Chapter 6. However, we end with a running total of 148 and no more rows to consider. We haven't yet reached 151. The quotient remains the same, $1 + 4 + 32 = 37$, but we're $151 - 148 = 3$ short, so the remainder is 3. Therefore, the scribe knew that there was a remainder because the final running total was less than the actual dividend.

Chapter 7

Exercise 1

```
         61                  923                 285                  85
   ┌──────────          ┌──────────          ┌─────────          ┌──────────
436)26776            104)96049             87)24874           276)23662
    26160                93600                17400               22080
    ─────                ─────                ─────               ─────
      616                2449                 7474                1582
      436                2080                 6960                1380
    ─────                ─────                ─────               ─────
      180                 369                  514                 202
                          312                  435
                         ─────                ─────
                           57                   79


        1137                  71                 392                 155
   ┌──────────          ┌──────────          ┌─────────          ┌──────────
 68)77328             992)71389            249)97731           555)86260
    68000                69440                74700               55500
    ─────                ─────                ─────               ─────
     9328                1949                23031               30760
     6800                 992                22410               27750
    ─────                ─────                ─────               ─────
     2528                 957                  621                3010
     2040                                      498                2775
    ─────                                    ─────               ─────
      488                                     123                 235
      476
    ─────
       12


        1575                 254                  67                  89
   ┌──────────          ┌──────────          ┌─────────          ┌──────────
 43)67737             279)70930            439)29685           751)67337
    43000                55800                26340               60080
    ─────                ─────                ─────               ─────
    24737                15130                 3345                7257
    21500                13950                 3073                6759
    ─────                ─────                ─────               ─────
     3237                 1180                  272                 498
     3010                 1116
    ─────                ─────
      227                   64
      215
    ─────
       12


          17                 237                   40                  74
   ┌──────────          ┌──────────          ┌─────────          ┌──────────
717)12510             388)92300            956)38548           466)34884
    7170                 77600                38240               32620
    ─────                ─────                ─────               ─────
    5340                14700                  308                2264
    5019                11640                                     1864
    ─────                ─────                                    ─────
     321                 3060                                      400
                         2716
                        ─────
                          344
```

Think About It

To find the primes less than 100, we use the sieve:

	2	3	4	5	6	7	8	9	10
11	12	13	14	15	16	17	18	19	20
21	22	23	24	25	26	27	28	29	30
31	32	33	34	35	36	37	38	39	40
41	42	43	44	45	46	47	48	49	50
51	52	53	54	55	56	57	58	59	60
61	62	63	64	65	66	67	68	69	70
71	72	73	74	75	76	77	78	79	80
81	82	83	84	85	86	87	88	89	90
91	92	93	94	95	96	97	98	99	100

which becomes, after erasing all multiples of 2, then 3, 5, 7, 11, 13, etc.,

	2	3	5	7	
11		13		17	19
		23			29
31				37	
41		43		47	
		53			59
61				67	
71		73			79
		83			89
				97	

Notice, 2 is the only even prime number.

Chapter 8

Exercise 1

$$\frac{6}{7} \div \frac{9}{6} = \frac{4}{7} \qquad \frac{6}{4} \times \frac{5}{6} = \frac{5}{4} \qquad \frac{7}{6} \times \frac{1}{3} = \frac{7}{18} \qquad \frac{9}{9} \div \frac{6}{2} = \frac{1}{3}$$

$$\frac{4}{3} + \frac{1}{2} = \frac{11}{6} \qquad \frac{4}{8} \times \frac{8}{7} = \frac{4}{7} \qquad \frac{3}{9} \div \frac{7}{3} = \frac{1}{7} \qquad \frac{6}{5} + \frac{5}{9} = \frac{79}{45}$$

$$\frac{6}{7} \div \frac{7}{2} = \frac{12}{49} \qquad \frac{6}{7} \div \frac{8}{2} = \frac{3}{14} \qquad \frac{2}{5} \div \frac{6}{2} = \frac{2}{15} \qquad \frac{5}{2} \div \frac{1}{9} = \frac{45}{2}$$

$$\frac{7}{2} - \frac{2}{8} = \frac{13}{4} \qquad \frac{7}{5} - \frac{6}{4} = -\frac{1}{10} \qquad \frac{6}{3} + \frac{5}{6} = \frac{17}{6} \qquad \frac{8}{8} \div \frac{2}{6} = 3$$

$$\frac{6}{8} - \frac{4}{5} = -\frac{1}{20} \qquad \frac{7}{9} \times \frac{2}{4} = \frac{7}{18} \qquad \frac{5}{8} + \frac{6}{2} = \frac{29}{8} \qquad \frac{3}{8} + \frac{8}{6} = \frac{41}{24}$$

$$\frac{9}{8} + \frac{8}{7} = \frac{127}{56} \qquad \frac{7}{7} - \frac{6}{8} = \frac{1}{4} \qquad \frac{6}{9} \div \frac{3}{6} = \frac{4}{3} \qquad \frac{2}{5} \times \frac{4}{9} = \frac{8}{45}$$

$$\frac{1}{8} + \frac{6}{2} = \frac{25}{8} \qquad \frac{8}{8} \div \frac{1}{6} = 6 \qquad \frac{2}{4} - \frac{8}{5} = -\frac{11}{10} \qquad \frac{3}{9} + \frac{4}{7} = \frac{19}{21}$$

$$\frac{5}{4} \div \frac{4}{7} = \frac{35}{16} \qquad \frac{5}{7} \div \frac{9}{8} = \frac{40}{63} \qquad \frac{4}{9} \times \frac{2}{8} = \frac{1}{9} \qquad \frac{6}{2} + \frac{2}{4} = \frac{7}{2}$$

$$\frac{4}{9} + \frac{3}{7} = \frac{55}{63} \qquad \frac{4}{2} + \frac{5}{5} = 3 \qquad \frac{3}{6} \div \frac{7}{2} = \frac{1}{7} \qquad \frac{1}{3} - \frac{4}{5} = -\frac{7}{15}$$

$$\frac{8}{3} - \frac{7}{8} = \frac{43}{24} \qquad \frac{3}{6} \times \frac{6}{3} = 1 \qquad \frac{2}{6} \div \frac{6}{3} = \frac{1}{6} \qquad \frac{7}{6} + \frac{3}{5} = \frac{53}{30}$$

Think About It

Here's the 6x6 table of possible outcomes for the sum of two 6-sided dice,

	1	2	3	4	5	6
1	2	3	4	5	6	7
2	3	4	5	6	7	8
3	4	5	6	7	8	9
4	5	6	7	8	9	10
5	6	7	8	9	10	11
6	7	8	9	10	11	12

Using the table above, we get the following showing each possible outcome, the number of times it appears in the table, and the associated probability,

Total	Count	Probability
2	1	$\frac{1}{36}$
3	2	$\frac{2}{36} = \frac{1}{18}$
4	3	$\frac{3}{36} = \frac{1}{12}$
5	4	$\frac{4}{36} = \frac{1}{9}$
6	5	$\frac{5}{36}$
7	6	$\frac{6}{36} = \frac{1}{6}$
8	5	$\frac{5}{36}$
9	4	$\frac{4}{36} = \frac{1}{9}$
10	3	$\frac{3}{36} = \frac{1}{12}$
11	2	$\frac{2}{36} = \frac{1}{18}$
12	1	$\frac{1}{36}$

The least likely outcomes are rolling a 2 or a 12, both with a probability of 1/36. The most likely outcome is a 7 with a probability of 1/6. The probability of rolling a 4 is 1/12. Therefore, to maximize your chance of winning, bet on 7. Of course, you'll still lose $1 - 1/6 = 5/6$ of the time, but that's better than losing $1 - 1/12 = 11/12$ of the time. As always with gambling, your best bet is not to bet at all; just keep your money.

Chapter 9

Exercise 1

$7^5 \times 7^7 = 7^{12}$ $\quad 7^8 \times 7^{-9} = 7^{-1}$ $\quad 4^4 \times 4^6 = 4^{10}$ $\quad 4^4 \div 4^5 = 4^{-1}$

$3^{-3} \times 3^{-2} = 3^{-5}$ $\quad 2^5 \times 2^{-7} = 2^{-2}$ $\quad 9^2 \div 9^{-5} = 9^7$ $\quad 3^7 \div 3^{-3} = 3^{10}$

$4^5 \times 4^9 = 4^{14}$ $\quad 5^{-5} \div 5^5 = 5^{-10}$ $\quad 2^{-7} \div 2^5 = 2^{-12}$ $\quad 3^4 \div 3^{-4} = 3^8$

$5^8 \times 5^{-9} = 5^{-1}$ $\quad 8^{-9} \div 8^{-9} = 8^0$ $\quad 2^7 \div 2^5 = 2^2$ $\quad 8^2 \div 8^9 = 8^{-7}$

$9^8 \times 9^{-8} = 9^0$ $\quad 5^5 \times 5^{-2} = 5^3$ $\quad 7^2 \div 7^2 = 7^0$ $\quad 8^8 \times 8^7 = 8^{15}$

$2^9 \div 2^{-5} = 2^{14}$ $\quad 3^6 \div 3^4 = 3^2$ $\quad 5^5 \div 5^{-8} = 5^{13}$ $\quad 6^3 \div 6^{-3} = 6^6$

$2^4 \div 2^5 = 2^{-1}$ $\quad 8^5 \times 8^{-2} = 8^3$ $\quad 7^2 \div 7^{-8} = 7^{10}$ $\quad 2^3 \times 2^3 = 2^6$

$6^5 \times 6^7 = 6^{12}$ $\quad 7^7 \div 7^6 = 7^1$ $\quad 2^8 \times 2^8 = 2^{16}$ $\quad 3^5 \times 3^{-7} = 3^{-2}$

$7^7 \times 7^{-6} = 7^1$ $\quad 4^3 \times 4^9 = 4^{12}$ $\quad 3^4 \times 3^5 = 3^9$ $\quad 2^5 \times 2^8 = 2^{13}$

$5^2 \div 5^{-5} = 5^7$ $\quad 2^4 \div 2^8 = 2^{-4}$ $\quad 6^8 \div 6^9 = 6^{-1}$ $\quad 2^{-2} \div 2^{-3} = 2^1$

Exercise 2
 a) $x = 13$
 b) $x = 100$
 c) $x = 3/4$
 d) $x = 1$

Think About It
The base 3 addition and multiplication tables:

+	0	1	2		×	0	1	2
0	0	1	2		0	0	0	0
1	1	**2**	10		1	0	1	**2**
2	2	**10**	**11**		2	0	2	**11**

Perhaps we should use base 3; look how simple the tables are!

The addition problem:

```
    21012
 +   2212
   ------
   101001
```

The multiplication problem:

```
     221
 ×    12
   ------
    1212
 +   221
   ------
   11122
```

Programmers are always confusing Halloween and Christmas because:

Oct 31 = Dec 25

This joke works because "Oct 31" means octal 31 (base 8) equals "Dec 25" (decimal, base 10). Let's see that this is so,

$$\begin{aligned}
\text{Oct } 31 &= 3 \times 8^1 + 1 \times 8^0 \\
&= 3 \times 8 + 1 \times 1 \\
&= 24 + 1 \\
&= 25
\end{aligned}$$

Remember, there are only 10 kinds of people in the world. Those who understand binary, and those who don't.

Chapter 10

Exercise 1
a) $\dfrac{1}{11} = 0.\overline{09}$ b) $\dfrac{5}{8} = 0.625$
c) $\dfrac{1}{13} = 0.\overline{076923}$ d) $\dfrac{5}{6} = 0.8\overline{3}$
e) $\dfrac{3}{7} = 0.\overline{428571}$ f) $\dfrac{7}{15} = 0.4\overline{6}$

Exercise 2
a) 97.005 > 97.004
b) 0.1011 > 0.01
c) 101.01 < 101.1
d) −0.123 < 0.123
e) 0 > −4.11
f) 0 < 4.11

Exercise 3

```
        -0.2502              0.5926            0.564              0.1091
    ×    0.067  ×           -0.0015×           -4.2     ×            6.1
         17514               29630             1128                 1091
        150120               59260            22560                65460
      -0.0167634          -0.00088890         -2.3688              0.66551


        -0.5397             679.8              0.0304             -686.1
    ×   -0.028   ×            4.2    ×        -0.0064×             0.0095
         43176              13596              1216                34305
        107940             271920             18240               617490
        0.0151116          2855.16          -0.00019456           -6.51795


          3.13              392.9             -40.19              -0.1201
    ×    -0.44    ×         0.0011   ×         0.0052   ×           -0.92
          1252               3929              8038                 2402
         12520              39290            200950               108090
        -1.3772             0.43219          -0.208988             0.110492
```

−241.1	430.7	−5.832	5.718
× −0.083	× −0.093	× −0.063	× 0.99
7233	12921	17496	51462
192880	387630	349920	514620
20.0113	−40.0551	0.367416	5.66082

Chapter 11

Exercise 1

2% of 255 is 5.1	49% of 228 is 111.72
76% of 91 is 69.16	82% of 67 is 54.94
15% of 607 is 91.05	55% of 94 is 51.7
10% of 312 is 31.2	24% of 74 is 17.76
79% of 55 is 43.45	32% of 733 is 234.56
10% of 69 is 6.9	62% of 139 is 86.18
71% of 424 is 301.04	78% of 740 is 577.2
1% of 59 is 0.59	26% of 483 is 125.58
57% of 510 is 290.7	27% of 600 is 162
4% of 989 is 39.56	15% of 853 is 127.95

Exercise 2

a) If all of something goes away, 100% of it is gone. Therefore, the percent change between 2020 and 1818 is a 100% decrease.

b) Popularity in 1930 was about 120 while popularity in 1945 was about 140, therefore, the percent change from 1930 is: $(120 - 140)/120 = -20/120 = -1/6 = 16.7\%$.

c) The percent change is: $(145 - 125)/145 = 20/145 = 4/29 = 13.8\%$.

d) Mary Shelley's novel, *Frankenstein*, was published in 1818 with a second edition published in 1831. The blip in the graph is due to initial interest in the novel.

e) Universal Studios released three Frankenstein movies in the 1930s: *Frankenstein* (1931), *The Bride of Frankenstein* (1935), and *Son of Frankenstein* (1939). The more recent increase in popularity is likely due to multiple factors: the 200th anniversary in 2018, the Dean Koontz *Frankenstein* series of books, and movies like *I, Frankenstein* (2014).

Frankenstein is considered by many to be the beginning of the science fiction era. Regardless, it is certainly an important, early work and well worth the read. You'll find both the 1818 and 1831 editions on Project Gutenberg, https://gutenberg.org/ebooks/41445 and https://gutenberg.org/ebooks/42324, respectively. Enjoy.

Think About It

Half-life	Time (millions)	^{14}C remaining (%)
0	0	100
1	1300	50
2	2600	25
3	3900	12.5
4	5200	6.25

The table shows time in millions of years. It takes 5.4 billion years for four full half-lives of ^{40}K, some 700 million years older than the Earth. In terms of half-lives, the Earth is about 4540/1300 = 3.49 half-lives old, leaving plenty of potassium-40 left for dating purposes.

The table shows the percent ^{40}K at the end of each half-life. These values are straightforward to find, just half the previous percentage. In general, however, exponential decay follows a formula like this,

$$\text{percent remaining} = e^{-kt}$$

for k a constant related to the half-life of the element, e a special number ($e = 2.718\ldots$), and t the time since decay began. For ^{40}K, k is approximately 0.0005 when t is given in terms of millions of years. Therefore, the percent of potassium-40 remaining in a sample as old as the Earth is roughly,

$$\text{percent remaining} = e^{-kt} = e^{-0.0005(4540)} = 0.103 = 10.3\%$$

We haven't discussed how to raise a non-integer to a non-integer power, like e raised to the $-kt$ power, but it is possible. The easiest way to do it is via the "y^x" key on a calculator.

Our discussion of *radiometric dating* has been necessarily simplified. Many other factors must be considered and accounted for to produce meaningful radiometric dates. Radiometric dating has transformed our understanding of history, the timing of the development of life, and our knowledge of the very age of the Earth itself. Before radiometric dating, paleontologists were aware of the sequence and the relative age of fossils, i.e., that one was older, perhaps much older, than another, but absolute age was still unknown. Radiometric dating provides those absolute ages.

Another recent example: in October 2021, researchers reported using extremely precise ^{14}C dating to determine that Vikings were living at the L'Anse aux Meadows site in Newfoundland, Canada, in 1021 AD, precisely 1000 years before the time I write this. Radiometric dating, like fezzes and bow ties, is cool.